Data Analytics for Smart Cities

Data Analytics Applications

Series Editor

Jay Liebowitz

Sport Business Analytics
Using Data to Increase Revenue and Improve Operational Efficiency
C. Keith Harrison, Scott Bukstein

Actionable Intelligence in Healthcare
Jay Liebowitz, Amanda Dawson

Data Analytics Applications in Latin America and Emerging Economies
Eduardo Rodriguez

Big Data and Analytics Applications in Government
Current Practices and Future Opportunities
Gregory Richards

Big Data Analytics in Cybersecurity
Onur Savas, Julia Deng

Data Analytics Applications in Education
Jan Vanthienen, Kristof De Witte

Intuition, Trust, and Analytics
Jay Liebowitz, Joanna Paliszkiewicz, Jerzy Gołuchowski

Research Analytics
Boosting University Productivity and Competitiveness Through Scientometrics
Francisco J. Cantu-Ortiz

Big Data in the Arts and Humanities: Theory and Practice
Giovanni Schiuma, Daniela Carlucci

Analytics and Knowledge Management
Suliman Hawamdeh, Hsia-Ching Chang

Data Analytics in Project Management
Edited by Seweryn Spalek

Data Analytics for Smart Cities
Edited by Amir H. Alavi, William G. Buttlar

For more information about this series, please visit: www.crcpress.com/Data-Analytics-Applications/book-series/CRCDATANAAPP

Data Analytics for Smart Cities

Edited by

Dr. Amir H. Alavi
Dr. William G. Buttlar

CRC Press is an imprint of the
Taylor & Francis Group, an **informa** business
AN AUERBACH BOOK

CRC Press
Taylor & Francis Group
6000 Broken Sound Parkway NW, Suite 300
Boca Raton, FL 33487-2742

First issued in paperback 2022

ISBN 13: 978-1-03-247596-7 (pbk)
ISBN 13: 978-1-138-30877-0 (hbk)

DOI: 10.1201/9780429434983

Publisher's Note
The publisher has gone to great lengths to ensure the quality of this reprint but points out that some imperfections in the original copies may be apparent.

Library of Congress Cataloging-in-Publication Data

Names: Alavi, Amir, editor. | Buttlar, William G., editor.
Title: Data analytics for smart cities / edited by Amir Alavi, William G. Buttlar.
Description: Boca Raton, Florida : CRC Press, Taylor & Francis Group, [2019]
| Includes bibliographical references and index.
Identifiers: LCCN 2018029000| ISBN 9781138308770 (hardback : alk. paper) |
ISBN 9780429434983 (e-book)
Subjects: LCSH: Smart cities. | Big data. | Quantitative research.
Classification: LCC TD159.4 .D38 2019 | DDC 307.760285--dc23
LC record available at https://lccn.loc.gov/2018029000

Visit the Taylor & Francis Web site at
http://www.taylorandfrancis.com

and the CRC Press Web site at
http://www.crcpress.com

Table of Contents

Preface

The development of smart cities will be one of the most important challenges over the next few decades. Governments and companies are leveraging billions of dollars in public and private funds for smart cities. Our next generation smart cities are heavily dependent on distributed smart sensing systems and devices to monitor the urban infrastructure. The smart sensor networks serve as autonomous intelligent nodes to measure a variety of physical or environmental parameters. They should react in time, establish automated control, and collect information for intelligent decision-making. In this context, one of the major tasks is to develop advanced frameworks for the interpretation of the huge amount of information provided by the emerging testing and monitoring systems.

Arguably, data analytics technologies play a key role in tackling this challenge. Data analytics applications involve collecting, integrating, and preparing time- and space-dependent data produced by sensors, complex engineered systems, and physical assets, followed by developing and testing analytical models to verify the accuracy of results. This is a multidisciplinary area including several paradigms such as machine learning, pattern recognition, statistics, intelligent databases, knowledge acquisition, data visualization, high performance computing, expert systems, etc. Moreover, it embraces the fairly new concept of "Big Data analytics" for the interpretation of smart cities' massive amounts of data.

This book brings together some of the most exciting new developments in the area of integrating advanced (Big) data analytics systems into smart cities along with complementary technological paradigms such as cloud computing and Internet of Things (IoT). This book serves as a reference for researchers and engineers in domains of advanced computation, optimization and data mining for smart civil infrastructure condition assessment, dynamic visualization, intelligent transportation systems (ITS), cyber-physical systems, and smart construction technologies. The chapters are presented in a hands-on manner to facilitate researchers in tackling similar applications.

Editors

Amir H. Alavi is an Assistant Professor at the University of Missouri (MU). His multi-disciplinary research integrates sensing, computation, control, networking, and information systems into the civil infrastructure to create cyber-physical infrastructure systems. Dr. Alavi's research interests include smart cities, smart infrastructure, structural health monitoring, deployment of advanced sensors, energy harvesting, and civil engineering system informatics. Dr. Alavi has authored four books and more than 150 publications in archival journals, book chapters and conference proceedings. Recently, he has been selected among the Scholar Google 300 Most Cited Authors in Civil Engineering, as well as Web of Science ESI's World Top 1% Scientific Minds.

William G. Buttlar is the Glen Barton Chair in Flexible Pavements at the University of Missouri (MU). He has over 100 peer-reviewed journal articles and nearly 300 total publications in the areas of pavements, materials, and smart infrastructure. Prior to joining the faculty at MU in 2016, he was a faculty member at the University of Illinois at Urbana-Champaign (UIUC) for 20 years, with 5 years of administrative experience serving as the Associate Dean of the UIUC Graduate College for Science and Engineering Programs, and Associate Dean of Graduate Programs for the College of Engineering. He was also the lead faculty member behind the establishment of City Digital at UILabs in Chicago.

Contributors

Amir H. Alavi
Department of Civil and Environmental Engineering
University of Missouri
Columbia, Missouri

Mohammed H. Almannaa
Charles E. Via, Jr. Dept. of Civil and Environmental Engineering
Center for Sustainable Mobility at the Virginia Tech Transportation Institute
Blacksburg, Virginia

Nasim Arbabzadeh
Department of Industrial and Systems Engineering
Rutgers, the State University of New Jersey
Piscataway, New Jersey

Huthaifa I. Ashqar
Charles E. Via, Jr. Dept. of Civil and Environmental Engineering
Center for Sustainable Mobility at the Virginia Tech Transportation Institute
Blacksburg, Virginia

Vahid Balali
Department of Civil Engineering and Construction Engineering Management
California State University, Long Beach
Los Angeles, California

William G. Buttlar
Department of Civil and Environmental Engineering
University of Missouri
Columbia, Missouri

Subasish Das
Associate Transportation Researcher
Texas A&M Transportation Institute
College Station, Texas

Mohammed Elhenawy
Center for Sustainable Mobility at the Virginia Tech Transportation Institute
Blacksburg, Virginia

Cheng Fan
College of Civil Engineering, Faculty of Engineering
Shenzhen University
Shenzhen, China

Ahmed Ghanem
Bradley Dept. of Electrical and Computer Engineering
Center for Sustainable Mobility at the Virginia Tech Transportation Institute
Blacksburg, Virginia

Fatemeh Golpayegani
The Future Cities Research Centre
Trinity College Dublin
Dublin, Ireland

Jie Gong
Dept. of Civil & Environmental Engineering
Rutgers, the State University of New Jersey
Piscataway, New Jersey

Xuan Hu
Dept. of Civil & Environmental Engineering
Rutgers, the State University of New Jersey
Piscataway, New Jersey

Jingke Hong
School of Construction Management and Real Estate
Chongqing University
Chongqing, China

Mohsen Jafari
Department of Industrial and Systems Engineering
Rutgers, the State University of New Jersey
Piscataway, New Jersey

Arash Jahangiri
Department of Civil, Construction, and Environmental Engineering
San Diego State University
San Diego, California

Mohammad Jalayer
Department of Civil and Environmental Engineering
Rowan University
Glassboro, New Jersey

Menglin S. Jin
Department of Atmospheric and Oceanic Science
University of Maryland at College Park
College Park, Maryland

Angela Li
NASA Goddard Earth Sciences Data and Information Services Center
Godard Space Flight Center
Greenbelt, Maryland

Clyde Zhengdao Li
College of Civil Engineering, Faculty of Engineering
Shenzhen University
Shenzhen, China

Jacqueline Liu
Urbana High School
Urbana, Maryland

Zhong Liu
NASA Goddard Earth Sciences Data and Information Services Center
Godard Space Flight Center
Greenbelt, Maryland

and

Center for Spatial Information Science and Systems
George Mason University
Fairfax, Virginia

Penousal Machado
CISUC, Department of Informatics Engineering
University of Coimbra
Coimbra, Portugal

Sahar Ghanipoor Machiani
Department of Civil, Construction, and Environmental Engineering
San Diego State University
San Diego, California

Abdollah Malekjafarian
School of Civil Engineering
University College Dublin
Dublin, Ireland

David Meyer
NASA Goddard Earth Sciences Data and Information Services Center
Godard Space Flight Center
Greenbelt, Maryland

Eugene J. OBrien
School of Civil Engineering
University College Dublin
Dublin, Ireland

William Teng
NASA Goddard Earth Sciences Data and Information Services Center
Godard Space Flight Center

and

Adnet Systems Inc.
Greenbelt, Maryland

Evgheni Polisciuc
CISUC, Department of Informatics Engineering
University of Coimbra
Coimbra, Portugal

Hesham A. Rakha
Charles E. Via, Jr. Dept. of Civil and Environmental Engineering
Bradley Dept. of Electrical and Computer Engineering
Center for Sustainable Mobility at the Virginia Tech Transportation Institute
Blacksburg, Virginia

Bruce Vollmer
NASA Goddard Earth Sciences Data and Information Services Center
Godard Space Flight Center
Greenbelt, Maryland

Bo Yu
College of Civil Engineering, Faculty of Engineering
Shenzhen University
Shenzhen, China

Huaguo Zhou
Department of Civil Engineering
Auburn University
Auburn, Alabama

1 Smartphone Technology Integrated with Machine Learning for Airport Pavement Condition Assessment

Amir H. Alavi and William G. Buttlar

CONTENTS

1.1 INTRODUCTION

America's airports are powerful economic engines and are an essential element of personal mobility and livable communities. They affect millions of jobs and impart trillions of dollars of impact on the American economy (ACI-NA 2010). Airports connect the nation's aviation system to other vital transportation modes (Hajek et al. 2011). The huge economic and social importance of airports imply the need to maintain condition and, when needed, expand capacity. One of the main steps in decision-making for maintenance and rehabilitation of airport pavements involves identifying and prioritizing future pavement preservation needs (Hajek et al. 2011). Toward this end, notable research has been conducted in the area of deployment of surveying and monitoring technologies for continuous assessment and safety evaluation of airside pavements (White 2014). In the U.S., the pavement condition index (PCI) surveying system is widely used for the assessment of the overall performance of airport pavement. The Federal Aviation Administration (FAA) is deploying analysis techniques for a more accurate prediction of PCI using historical data (Hall et al. 2014). A number of researchers have proposed new approaches based on the PCI system to provide more insight into pavement failure modes (Parsons and Pullen 2014; Butt and Pemmaraju 2014). Other approaches such as high-resolution surface scanning, ground penetration radar (GPR), and falling weight deflectometer (FWD) testing are utilized by authorities for structural health monitoring of airfield pavements (Adcock et al. 1995; Arraigada et al. 2009; Duan et al. 2011). Pavement instrumentation is another technology that allows more reliable and cost-effective field measurement of service conditions, loads, and stresses (Yang et al. 2014; Rabaiotti et al. 2017). In general, day-to-day collection of information using the aforementioned surveying and monitoring methods is costly and requires significant manpower. Furthermore, technical difficulties and economic issues associated with installing and maintaining instrumentation have hampered their large-scale implementation in pavement systems. In addition, pavement inspection and maintenance activities should be kept to a minimum to avoid undue disruption to airport operations (White 2014). Therefore, there is a need for simple, efficient, and low-cost tools to facilitate frequent monitoring and evaluation of airport pavements.

Now owned by a majority of U.S. citizens, powerful, sensor-packed smartphones can be used as mobile sensing units in "smart" civil infrastructure monitoring systems (Ozer 2016). These sensors include barometers, gyroscopes, accelerometers, proximity sensors, cameras, touch screens, microphones, ambient light sensors, and magnetometers. Smartphones now possess significant onboard memory and computing capabilities, opening the door for efficient cloud computing schemes involving edge-of-cloud computing and reduced bandwidth data transmission. Smartphones are built upon mobile operating systems and wireless communication hardware that can be used for field data collection and uploading of real-time data to cloud servers via Bluetooth, Wi-Fi, 3G, 4G, and 5G networks. All of these features imply that smartphones can become central to future civil infrastructure monitoring systems. In the last few years, there has been a surge of research on mobile sensing paradigm for data collection, signal processing, and data visualization in practical real-world applications. Today's smartphones are potentially useful tools for pavement condition assessment in a cost-efficient manner with large spatial coverage. In addition, they provide an opportunity for frequent, comprehensive, and quantitative monitoring of pavement infrastructure. In recent years, many studies have been conducted to explore the feasibility of using smartphones to estimate pavement condition. In general, pavement condition can be classified by the defects in the pavement surface that adversely affects the ride quality of vehicles. These anomalies may be in the form of surface roughness, unevenness, potholes, cracks, deterioration or other damage (Douangphachanh and Oneyama 2013; Islam and Buttlar 2012).

This study explores the use of smartphones for the assessment of airport pavement condition. This low-cost and efficient monitoring technology can be of particular importance for regional airports that may not have sufficient resources for surveying and monitoring at the desired frequency. A smartphone application was used to collect acceleration data for airport pavements at the 26 state funded regional airports in Missouri. The acceleration data was then used to calculate international roughness index (IRI) values. The obtained IRI values were found to be in good agreement with the construction and maintenance records of the airports. In addition, a prediction model was developed for PCI in terms of the smartphone-measured IRI using a machine learning method called genetic programming (GP).

1.2 SMARTPHONE-DRIVEN ASSESSMENT OF AIRPORT PAVEMENT CONDITION

This research builds upon previous work conducted at the University of Illinois at Urbana-Champaign (Islam et al. 2014, 2016; Islam 2015; Stribling et al. 2016) and presents the results of ongoing research at the University of Missouri-Columbia (MU) on the application of smartphones for measurement of pavement roughness and condition assessment. The first phase of the study focused on the calibration of the smartphone-based monitoring technology for the estimation of IRI on GAA pavements across Missouri. Although less commonly used on airport pavements as compared to highways, IRI is an internationally accepted pavement condition indicator because of its correlation to ride quality, user delay costs, maintenance costs, and fuel consumption (Islam and Buttlar 2012). To calculate IRI, pavement profile is back-estimated from vehicle cab acceleration data recorded by an Android-based smartphone application using an inverse state space model. The model considers the physics of the mass-spring-damper system of the vehicle's sprung mass. (More details about this model can be found in Islam et al. 2014; Islam 2015; Stribling et al. 2016).

The main vehicle parameter in the inverse state space model are curb weight, sprung mass (m_1), unsprung mass (m_2), suspension spring (k_1), tire spring (k_2), dashpot (c_1), and dampening coefficient (ζ). Curb weight, m_1, m_2, k_1, and k_2 are usually published by vehicle manufacturers. The coefficient ζ of typical passenger vehicles ranges from 0.200 to 0.400 and can be calculated as $c_1 / 2\sqrt{m_1 \times k_1}$ (Aisopoulos 2011; Stribling et al. 2016). The analyses were performed using a MATLAB® script to calculate the IRI values. In order to improve assessment repeatability for the purposes of the current research, only one smartphone model (Samsung Galaxy S8), one type of smartphone car mount and one vehicle type (SUV) was used for data collection through the entire project. The smartphone-measured IRI values were obtained for test roads near Columbia, MO and compared with known IRI values measured by Missouri Department of Transportation's (MoDOT) automatic road analyzer (ARAN) van; the values are stored in MoDOT's Transportation Management Systems (TMS) database. As previously mentioned, PCI surveying system is widely used in U.S. for the assessment of airport pavement overall performance. Therefore, an extensive survey was carried out to obtain the PCI values for the 26 genetic algorithm (GA) airports in Missouri. Then, a machine learning model was developed to find the functional relationship between the PCI of the airports and the smartphone-based IRI.

1.2.1 Description of Smartphone Application

The roughness capture application, developed by Applied Research Associates (ARA), was used to collect acceleration in three orthogonal directions, a timestamp and GPS coordinates, and to store them in an ASCII text file. The data collection rate is specified by the user, generally in the range of 10 to 140 samples per second. Higher sampling rates are possible depending upon smartphone hardware. In general, the higher the data collection rate, the better the accuracy of the estimated pavement profile (Islam 2015). In this study, the data collection from the application was set to 7 milliseconds per data point or approximately 142 data points per second.

1.2.2 Smartphone Characteristics

Measurements showed that a maximum of about 135 points/second can be reliably obtained from the cellphone (Samsung Galaxy S8) used in this study. For the standard speed of 50 mph, the vehicle travels 880 inches/second. Thus, the spacing of acceleration data points is 6.52 inches. The application can collect localization information either from the internal GPS or from a cellular network. The measurement type may also be specified as acceleration only, gravity only or gravity and acceleration. Roughness is mostly influenced by the wavelength ranging from 4 to 100 feet (1.23 to 30.48 m), whereas maximum sensitivity resides in the range of 8 to 51 feet (2.46 to 15.54 m) because of the high gain for profile slope (Islam et al. 2015). Therefore, both low-pass and high-pass filters have been utilized to remove wavelengths greater than 100 feet (30.48 m) and less than 4 feet (1.22 m), respectively, from the acceleration data. Roughness was estimated in terms of IRI of each 0.1-mile section.

1.3 CASE STUDY OF MISSOURI AIRPORTS

To investigate the potential for the smartphone application to be used as a tool for assessing airport pavement condition, a case study was undertaken for Missouri state funded general aviation airports. The following sections describe the case study, including the calibration process, data collection methodology, and the results.

1.3.1 Calibration Study

A calibration study to find the optimal model parameters was first performed on a selected section of MO-10E. The test location is shown in Figure 1.1. The corresponding IRI values measured by the ARAN van were obtained from MoDOT's transportation management system (TMS) database. The ARAN-based IRI measurements were taken in April 2016. The ARAN-based IRI values were calculated at speed of 50 mph (80 km/hr). The smartphone test runs were also conducted at a speed of 50 mph. The vehicle used for both the MO-10E calibration and the airport data collection was a 2009 Chevy Traverse. The vehicle suspension and smartphone parameter settings from the calibration are shown in Table 1.1. The IRI values were calculated for three ζ values (0.2, 0.3, 0.4) to find the best match with the ARAN data. The test section was selected based on its proximity to the first monitored airport (Excelsior Springs airport) and the availability of ARAN data in TMS. The characteristics of the MO-10E test section are as follows:

- Road: Excelsior Springs, MO-10 E, Travelway Id: 5015
- Start: Log 40.449 (Coordinates: 39.3350614, −94.1341107)
- End Log: 43.74 (Coordinates: 39.3406533, −94.192451)

Each test condition was replicated six times to test the repeatability of the system and to arrive at a reliable average. The Android-based smartphone was positioned horizontally on the vehicle dashboard (Figure 1.2). The best calibration results obtained for $\zeta = 0.4$ ($c_1 = 5261$ Ns/m) are shown in Figure 1.3. The performance measures of correlation coefficient (R), root mean squared error (RMSE), and mean absolute error (MAE) were then calculated for the data.

Figure 1.1 Street view of MO-10E test section (Google).

Table 1.1: Vehicle Suspension and Smartphone Parameter Settings for GA Airport Evaluation

	Parameter	Value
Vehicle	Make/Model/Year	Chevrolet Traverse LE 2009
	Curb Weight	2298 kg
	Sprung Mass, m_1	724 kg
	Unsprung Mass, m_2	80 kg
	Suspension Spring, k_1	71008 N/m
	Dampening Coeff., ζ	0.2, 0.3, 0.4
	Dashpot, c_1	2868, 4302, 5735 Ns/m
	Tire Spring, k_2	80000 N/m
Smartphone	Model	Samsung Galaxy S8
	Localization (GPS, Cellular network)	GPS
	Measurement type (acceleration, gravity, gravity and acceleration)	Acceleration
	Collection Rate	7 milliseconds per data point ($\approx$ 142 data points per second)

Figure 1.2 Smartphone mounting arrangement on top of vehicle dashboard.

As illustrated in Figure 1.3, the averaged IRI values measured by the smartphone system were in good agreement with the ARAN-measured IRI. The increased suspension dampening seems to help provide more consistency across the test runs. According to Sayers et al. (1986), the suspension characteristics of a vehicle is the single most important factor in measuring IRI. A vehicle possessing a softer suspension will oscillate with a longer wavelength than one with a stiffer suspension. This difference in oscillation can alter the perceived roughness in a road when measuring accelerations in the vehicle cab, sometimes referred to as the sprung mass (Stribling et al. 2016). The smartphone-based roughness system is also assessed in terms of its ability to classify pavement

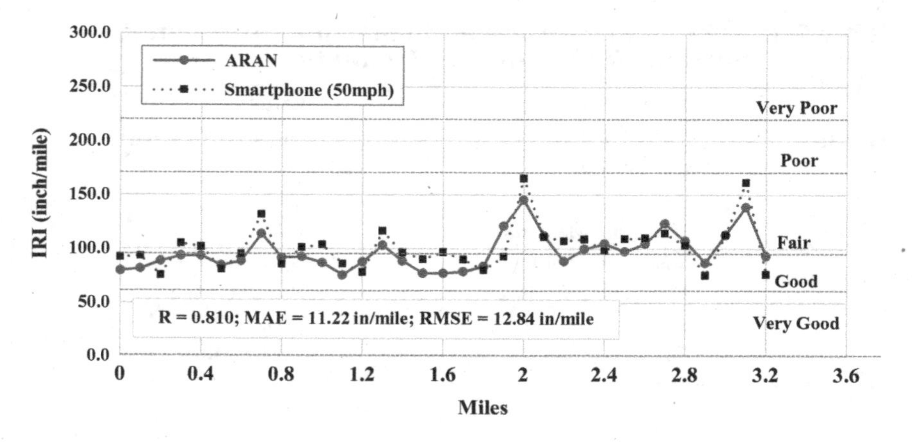

Figure 1.3 Smartphone calibration on MO-10 (1 inch/mile = 0.016 m/km).

according to Moving Ahead for Progress in the 21st Century Act (MAP-21) criteria. MAP-21 requires the states to provide pavement IRI data for every 0.1-mile pavement section for the interstate and non-interstate highway systems annually and biannually, respectively. Pavement ride quality can be categorized into five groups (U.S. Department of Transportation 2000), as shown in Table 1.2. In Figure 1.3, the right vertical axis has been labeled according to MAP-21 smoothness criteria threshold values. As seen, the smartphone-based IRI assessment system can categorize pavement condition based on roughness accurately for most of the segments (28 out of 33 were correctly categorized). Therefore, $\zeta = 0.4$ and the other parameters shown in Table 1.1 were used during the data collection process for each of the 26 Missouri airports in this study.

1.3.2 Missouri Airport Smartphone Data Collection Methodology

The roughness data for the Missouri state funded GAAs were collected using a 2009 Chevy Traverse and Samsung Galaxy S8. The IRI values are reported for the right, centerline, and left lanes of the 26 airfield pavements. Right and left lanes refer to assessments made with an offset of around 10 to 15 feet from the centerline. The test runs were conducted at two different speeds (+/−2 mph): 30 mph (48 km/hr) and 40 mph (64 km/hr) and in triplicate replication. The IRI

Table 1.2: Pavement Ride Quality for Interstate and Non-Interstate Facilities Based on Roughness (U.S. Department of Transportation 2000)

	IRI Rating (inch/mile)[a]		
Category	Interstate	Non-Interstate	Interstate and NHS Ride Quality
Very Good	< 60	< 60	Acceptable 0–170
Good	60–94	60–94	
Fair	95–119	95–170	
Poor	120–170	171–220	Less than acceptable > 170
Very Poor	> 170	> 220	

[a] 1 inch/mile = 0.016 m/km

values were calculated for each 0.1-mile using the smartphone acceleration data and a MATLAB script. Vehicle suspension and smartphone parameter settings are presented in Table 1.1.

1.3.3 Missouri Airport Smartphone Data Collection Results for Each Airport

The estimated average IRI values from the smartphone data collection for some of the selected airport are shown Figure 1.4. Figures 1.5 and 1.6 show the IRI values for the Missouri state-funded GAAs based on their pavement roughness for 30 and 40 mph speeds, respectively. The results for these two speeds are very similar. However, the preliminary measurements showed that the speed of 30 mph provides more consistent results than 40 mph and therefore, it was used for further analysis. This is also a safer speed for data collection in GA airports. Based on the results for 30 mph runs, the classification of ride quality for the airfield pavements investigated is as follows:

- Very Good: Van Buren
- Good: Branson, Carrollton, Mississippi County, Doniphan, Clarkton, Hornersville, El Dorado Springs, Unionville, Excelsior Springs, Willow Springs
- Fair: Mount Vernon, Gideon, Thayer, Versailles, Richland, Steele, Ava, Buffalo, Rhineland, Albany, Bismarck, Stockton, Monroe City
- Poor: Mansfield
- Very Poor: Bonne Terre

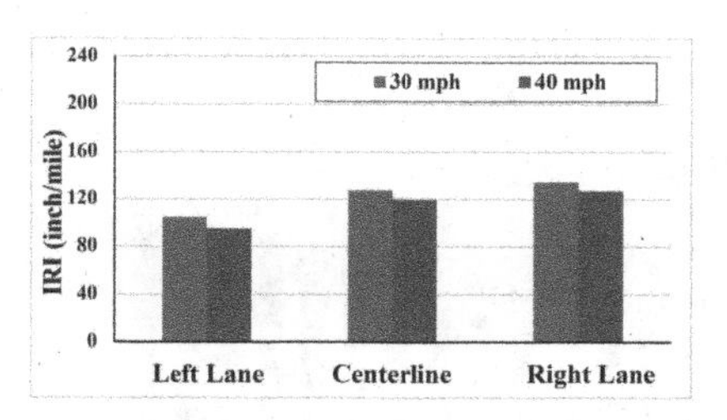

(a) Albany Municipal Airport

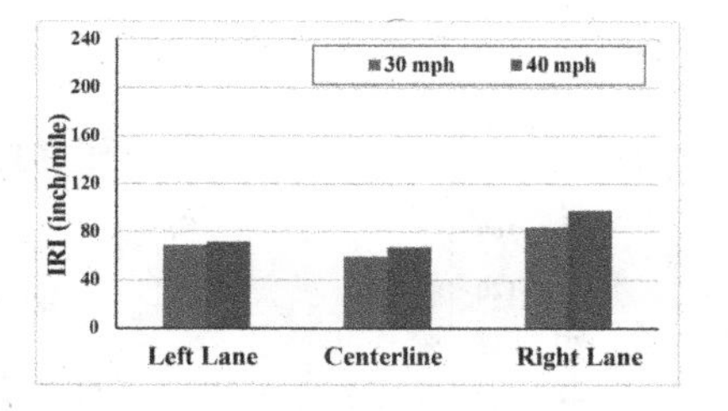

(b) Bollinger-Crass Memorial Airport

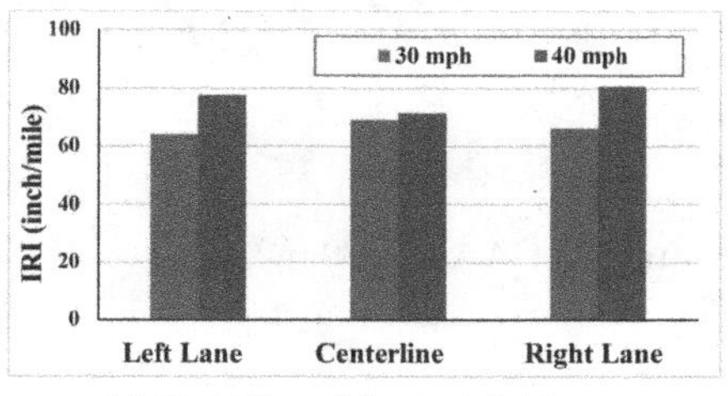

(c) Carrollton Memorial Airport

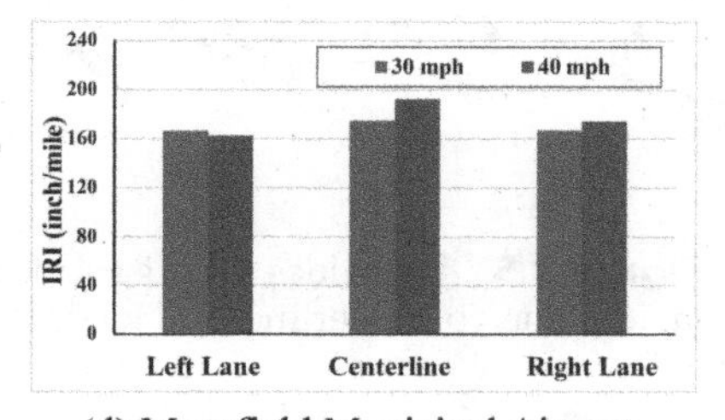

(d) Mansfield Municipal Airport

Figure 1.4 Estimated average IRI values for the Missouri airports. (a) Albany Municipal Airport. (b) Bollinger-Crass Memorial Airport. (c) Carrollton Memorial Airport. (d) Mansfield Municipal Airport.

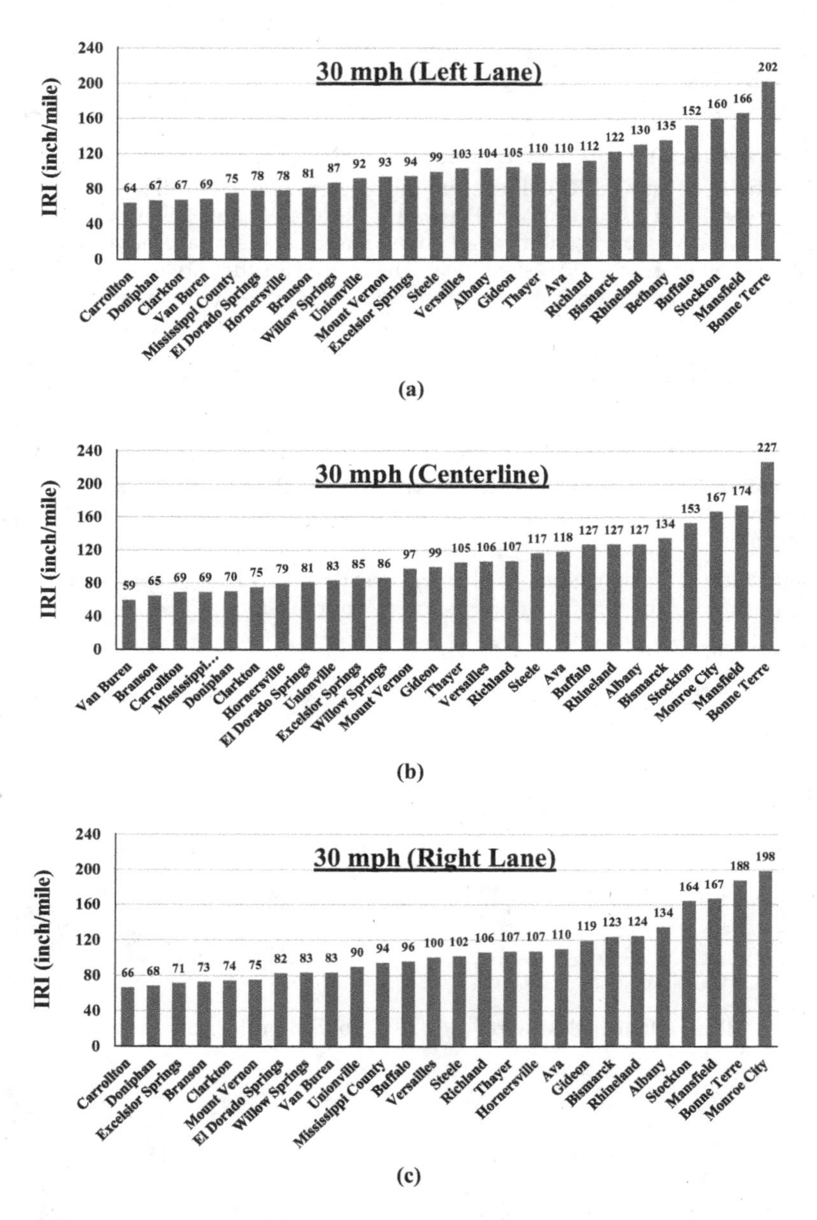

Figure 1.5 IRI values for Missouri state-funded GAAs for 30 mph speed: (a) left lane, (b) centerline, and (c) right lane.

The results are in good agreement with the construction and maintenance records for each airport as shown in Table 1.3. In this table, the airports are ordered based on centerline smoothness for 30 mph speed. As seen in Table 1.3, the Bonne Terre Municipal Airport pavement is the least smooth. Although this airport was constructed in 1966, there are no records of pavement maintenance

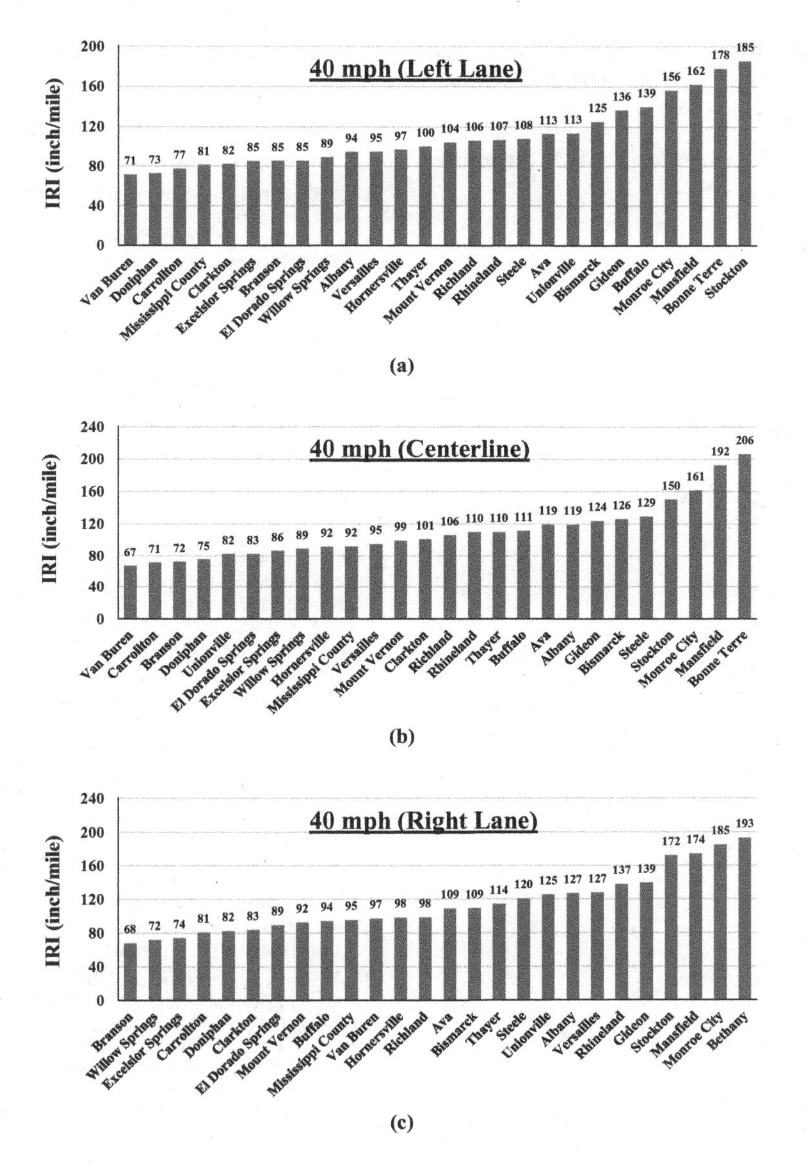

Figure 1.6 IRI values for Missouri state-funded GAAs for 40 mph speed: (a) left lane, (b) centerline, and (c) right lane.

operations, which concurs with the visual condition of the runway, which is poor. The lowest IRI belonged to the Bollinger-Crass Memorial Airport (Van Buren, MO), which was constructed in 1971, sealed in 1985 and 1987, and rehabilitated between 2012 and 2013. The acceptable correlation between the smoothness results and the construction and maintenance records demonstrates that the smartphone application has the potential to be an effective tool for evaluating airport pavement condition at GA airports.

Table 1.3: Construction and Maintenance Records for the Missouri Airports

No.	Airport	Smartphone-based IRI (inch/mile)	Construction and Maintenance Records
1	Van Buren	59	1971: Original construction (runway, taxiway, apron) 1985: Surface seal (runway, taxiway) 1997: Surface seal (runway, taxiway, apron) 2012/2013: Pavement Rehab (runway, connecting taxiway)
2	Branson	65	1971: Original construction (runway, connecting taxiway, parallel taxiway, apron) 2009: Maintenance (runway, taxiway), Seal coat, crack repair, and isolated pavement removal and replacement (apron, taxiway) 2013: Pavement removal (runway, connecting taxiway, apron)
3	Carrollton	69	1963: Originally paved 1968: Overlay (runway, taxiway), Expansion (apron) 1980: Pavement rehab (runway, taxiway) 1984: Crack seal (runway, taxiway) 1986: Asphalt reconstruction (apron) 2004: Crack seal (runway, taxiway) 2009: Pavement Maintenance: Crack seal, clean joints, patching (runway)
4	Mississippi County	69	1973: Original construction (runway, taxiway, apron) 1998: Repair and seal (runway, taxiway, apron) 7/18/2012: Bituminous overlay and crack seal (runway, taxiway, turnaround, apron), Additional reconstruction (apron)
5	Doniphan	70	1989: Original construction 1991: Seal (runway, taxiway) 1992: Unknown construction/maintenance (no detailed records) 1994: Resurfacing (runway) 1996: Overlay (runway) 2009: Seal coat and crack seal (runway, connecting taxiways 1 and 2, taxiway 3c)
6	Clarkton	75	1967: Runway constructed 2007: Pavement Removal, Replacement, 6″ crushed aggregate base, 3″ pavement layer, 1.75″ overlay, crack seal (runway, taxiway), Pavement removal, replacement, 6″ crushed aggregate base, 3″ pavement payer, 1.75″ overlay, 1.75″ cold mill, crack seal (apron, hanger)
7	Hornersville	79	*Initial construction date unknown 2009: Slurry seal and expansion (apron), removal of existing angled taxi lane, and rigid pavement repair and crack sealing (taxiway and apron) 2010: Seal coat, pavement removal, 6″ asphalt overlay, crack seal (apron and runway)
8	El Dorado Springs	81	* Initial construction date unknown 1982: Construct apron, reconstruct taxiway 1983: Seal runway 1999: Reconstruct apron, runway, and taxiway 2005: Seal runway, taxiway, and apron 2013: Runway and taxiway seal coat and apron reconstruction

(Continued)

Table 1.3: (Continued) Construction and Maintenance Records for the Missouri Airports

No.	Airport	Smartphone-based IRI (inch/mile)	Construction and Maintenance Records
9	Unionville	83	*Initial construction date unknown 10/3/2012: Crack repair, full-depth bituminous patching, bituminous overlay (runway, connecting taxiway, apron)
10	Excelsior Springs	85	1952: Original construction date 1986: Seal runway and taxiway 1987: Pave runway and taxiway 2015: Reconstruct runway, seal coat, and crack seal (apron, connecting taxiway)
11	Willow Springs	86	*Initial construction date unknown 2006: Rehabilitation (widening and 2″ overlay, new pavement construction) (runway) 2010: Seal coat (runway, taxiways, aprons)
12	Mount Vernon	97	*Initial construction date unknown 2002: 2″ bituminous overlay (runway, taxiway, aprons) 2009: Routing/Crack Seal and Seal Coat (runway, taxiway, apron, and south turnaround) (About 4000 linear feet).
13	Gideon	99	1942: Original construction date 1960s: Some maintenance and overlay (details unknown) 1970s: Some maintenance and overlay (details unknown) 1990s: Some patchwork (details unknown) 2000s: Some patchwork (details unknown) 2013: Mill and overlay (runway). Alternative 1: overlay of parallel taxiway and connecting taxiway (existing footprint). Alternative 2: overlay of parallel taxiway and connecting taxiway (25′ width).
14	Thayer	105	*Initial construction date unknown 2012: Seal coat (runway, taxiway)
15	Versailles	106	1970: Original construction 1981: Overlay (runway, taxiway, apron) 1986: Seal Coat runway and taxiway 1995: Seal Coat runway, taxiway, and apron resurfacing (001-003 assumed, no documentation) 2006: Overlay (runway, connecting taxiway), Seal apron, rehab hanger 2007: Joint and crack repair, some full-depth, some pavement removal and overlay (Connecting taxiway and aprons). Joint and crack repair, Petromat 2″ overlay (Runway). Seal coat (TLA-001). Reconstruct with PCC (TLA-002/03). 2009: Seal runway, connecting taxiway 2014: Crack and joint seal (runway, connecting taxiway, and TLA), Seal coat (runway), reconstruct pavement on apron. 2015: Pavement removal, crack seal, seal coat (location unknown)

(Continued)

Table 1.3: (Continued) Construction and Maintenance Records for the Missouri Airports

No.	Airport	Smartphone-based IRI (inch/mile)	Construction and Maintenance Records
16	Richland	107	1970: Airport constructed 1985: Pavement repair (runway, taxiway, and apron) 2002: 2" bituminous overlay (runway, taxiway, and apron) 2010: Seal coat, crack repair (runway, taxiway, apron)
17	Steele	117	1944: Original construction date 1973: Overlay runway (date approximate) 1985: Possible overlay (date approximate, unable to confirm) 1995: Possible seal coat (date approximate, unable to confirm). Per airport personnel, generally try to seal coat every 10 years but unsure of dates prior to 2005. 2005: Seal coat, crack repair (runway, taxiway, apron) 2007: Widened runway, removal of existing deteriorated pavements 2013: Crack seal, seal coat (runway, taxiway, apron)
18	Ava	118	1967: Apron fencing constructed, no documents 1974: Runway extended, no documents 1979: Mill and Overlay (runway, connecting taxiway, TLA, RTA), apron expansion, no documents 1984: Seal coat (runway, taxiway, apron), no documents 1989: Leveling overlay course on runway, no documents 1994: Seal coat (runway, taxiway, apron), no documents 2006: Crack and joint sealing, seal coat (runway) 2015: Hanger taxiway reconstruction, no documents
19	Buffalo	127	1952: Airport constructed 1988: Runway resurfacing, no documents 1997: Pavement rehab (runway, connecting taxiway, apron) 2004: Seal coat (runway, connecting taxiway, apron) 2013: Seal coat, crack seal, and repair (runway, connecting taxiway, heliport)
20	Rhineland	127	1974: Constructed (Phase 1) 1981: Constructed (Phase 2) 1986: Seal Coat (runway and taxiway) 1988: Seal Coat (runway and taxiway) 1993: Seal Coat (runway) 1995: Extensive flood damage, required overlay and replacement of damage with new surface 2001: Reconstruction and, expansion(apron and taxiway) 2002: Seal Coat (runway and connecting taxiway) 2014: General maintenance: crack and joint seal, seal coat (runway, turnarounds, connecting taxiway) /2015: Clean and seal joints, seal coat (runway, taxiway, turnarounds)

(*Continued*)

Table 1.3: (Continued) Construction and Maintenance Records for the Missouri Airports

No.	Airport	Smartphone-based IRI (inch/mile)	Construction and Maintenance Records
21	Albany	127	1982: Runway paved (original construction), no documents 1987: Pavement rehab: 5" PCC Runway Overlay and PCC Panel repairs, apron expansion 1990: Expansion (runway) 2008: 5" PCC Runway: crack and joint sealing, 12 panel repairs 2010: Crack repair, panel replacement, partial depth patching on runway
22	Bismarck	134	1965: Runway originally constructed 1982: Seal Coat (runway) 1999: Repair/Resurface, no documents (runway, taxiway, apron) 2008: Seal Coat (runway, taxiway, apron seal coat), no documents 2010: Crack and joint seal, and friction surface seal (runway, connecting taxiway, apron)
23	Stockton	153	1964: Airport is constructed 2007: Slurry sealing, crack and joint sealing, sealcoat surface (apron) 3/29/2009: Full-depth pavement repair, crack and joint sealing, seal coat treatment (runway and taxiway)
24	Monroe City	167	*Initial construction date unknown 1979: Extend/expand runway 2006: Crack and joint seal, seal coat (runway, taxiways, aprons) 2010: Full-depth, crack repair, and seal coat (runways, taxiways, aprons)
25	Mansfield	174	*Initial construction date unknown 2010: Seal coat and crack seal (runway, taxiway, apron) 2013: Crack seal, seal coat (runway, taxiway, apron)
26	Bonne Terre	227	1966: Runway paved (Lead Belt Materials, 2") 1966 - Present: No pavement maintenance, occasional crack sealing and weed killer in cracks

1.3.4 Discussion

While the smartphone-based IRI roughness results appear to be quite reasonable, there are several issues that should be addressed in future research:

- During the measurement at airports with rough pavements (e.g., Bonne Terre Airport), it was observed that the data had outliers. The outliers show the significant effect of vehicle wander on collecting pavement roughness given that all other conditions were constant (with weather/temperature being relatively the same). However, the effect of vehicle wander can be overcome by collecting and averaging larger volumes of data. It is recommended that at least six replications be performed for each section. In general, further validation should be done for very rough pavement sections. In addition, the current android application does not automatically eliminate outliers in the data nor does it conduct any onboard analytics. These features can be added to the application along with real-time estimation of IRI.

- The smartphone application used in this study collected about 135 acceleration points per second. The vehicle running at 50 mph travels 880 inches per second, resulting in spatial distance between acceleration data points of 6.52 inches. Therefore, the smartphone application may very likely be missing peak accelerations due to the relatively slow data collection rate. Unlike the smartphones, the inertial profilers have a very high sampling rate (1 kHz). However, with the expected advancement of smartphone technology, higher data collection rates will be possible, potentially rendering IRI estimates on rough pavements even more accurate. Another idea is to attach commercially available accelerometers to a smartphone and collect data at a higher sampling rate. This should make the resulting assessments more accurate and repeatable.
- A simple trial-and-error method was used in this study to calibrate the vehicle suspension parameters. An improved calibration method could involve the development of a robust algorithm to optimize vehicle suspension parameters and minimize the differences between IRI values measured from the smartphone and ARAN. An ideal implementation would include upscaling to a crowd-sourced data collection scheme.
- For future research, measurements could be taken using smartphones placed in general aviation aircraft in order to capture aircraft cab acceleration data and thereby generate a more direct assessment of the impacts of pavement roughness on aircraft wear-and-tear and pilot/passenger comfort and safety.
- It is also possible to extract some additional parameters directly from the discrete acceleration data and develop algorithms for PCI estimation that are even more precise than the IRI-only prediction of PCI. For instance, we already know that the smartphone acceleration trace can identify joints in PCC and larger cracks in asphalt. It seems feasible to characterize asphalt cracks of high severity that are related to thermal cracks, reflective cracks, block cracks and other linear cracking. Since many of these (if severe) would lead to high PCI deductions (but not necessarily lending directly to high IRI), adding the number and magnitude of these discrete events to a prediction algorithm would make it more accurate, and would assist in the development of new rating parameters tailored for airports.

1.4 PREDICTION OF PCI BASED ON SMARTPHONE-MEASURED IRI

The main goal of this phase of research was to estimate PCI from smartphone-measured IRI using a powerful machine learning technique – GP. The PCI values for the 26 airfield pavements in Missouri were obtained by conducting the industry standard, foot-on-ground PCI survey. This was performed by the University of Missouri and ARA, following ASTM D5340-12 (2012).

1.4.1 Machine Learning Method

Machine learning (ML) techniques are considered as powerful, modern alternative to traditional analysis methods for predicting the behavior of real-world systems. They automatically learn from experience and extract various discriminators in the process (Mitchell 1997). ML has a range of well-known branches, such as the artificial neural network (ANN), fuzzy inference system (FIS), adaptive neuro-fuzzy system (ANFIS), and support vector machines (SVM). These techniques have been successfully deployed to solve problems in the engineering field (e.g., Patra 1997; Zang and Imregun 2001). Among different ML techniques, ANNs have been widely used in the field of civil infrastructure condition assessment. Despite their good performance, ANNs are considered as black-box models. That is, they are not capable of generating practical prediction equations. Another limitation of ANNs is that their structure should be defined in advance (Alavi and Gandomi 2011; Alavi et al. 2016).

Inspired by the natural evolution and the Darwinian concept of "survival of the fittest," evolutionary computational (EC) methods are well-known branches of soft computing. Some of the subsets of EC are evolutionary strategies (ESs) (Schwefel 1975) and evolutionary programming (EP) (Fogel et al. 1996). These techniques are collectively known as evolutionary algorithms (EAs). In general, an EA consists of an initial population of random individuals improved by a set of genetic operators (e.g., reproduction, mutation, and recombination). The individuals are encoded solutions in form of binary strings of numbers evaluated by some fitness functions (Coello et al. 2007). Improvement of the population is a process to reach the fittest solution with the maximum convergence. Typically in an EA, a population of individual is randomly created and then the members are ranked according to a fitness function. The members with the highest fitness ranking are given a higher chance to become parents for the next generation (offspring). The approach used to generate offspring from the parents is referred to as the reproduction heuristic. Then selected members are randomly transformed into new members via mutation, recombination or crossover. These steps are repeated until the convergence conditions are satisfied and the fittest member is selected (Koza 1992; Coello et al. 2007). The differences between EAs are in the way that they represent the individual structures, types of selection mechanism, forms of genetic operators, and measures of performance (Alavi et al. 2016).

The GA technique has been shown to be a robust EA for dealing with a wide variety of complex civil engineering problems (Unal et al. 2014; Momeni et al. 2014). GP is a specialization of GA where the encoded solutions (individuals) are computer programs rather than binary strings (Koza 1992; Banzhaf et al. 1998). Figure 1.7 shows a comparison of the encoded solutions (individuals) by GA and GP (Alavi et al. 2016). In GP, inputs and corresponding output data samples are known and the main goal is to generate predictive models relating them (Weise 2009). GP has several advantages over the other soft computing techniques such as ANNs. A notable feature of GP and its variants is that they can produce highly nonlinear prediction equations without a need to pre-define the form of the

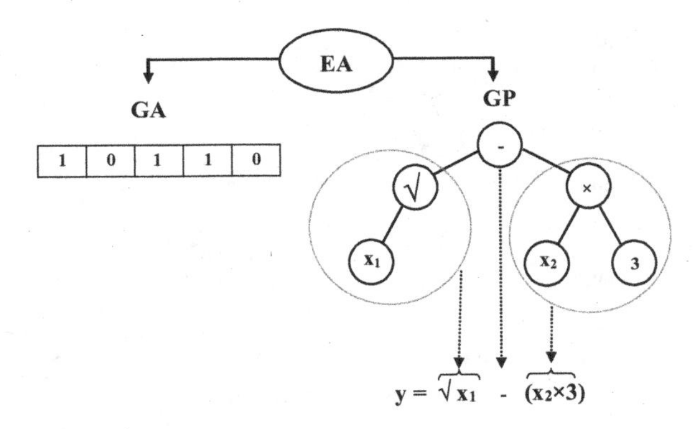

Figure 1.7 A comparative illustration of encoded solutions by GA and GP.

existing relationship (Alavi and Gandomi 2011). However, the application of GP and its variants to infrastructure condition assessment has yet to be fully exploited.

Different variants of GP have been proposed to improve the traditional GP algorithm (Weise 2009). The tree-shaped variant is the mostly widely used representation of GP algorithms. In addition to the classical tree-based GP, there are other types of GP approaches, i.e., linear and graph-based (Banzhaf et al. 1998). The emphasis of the present study was placed on linear-based GP techniques. The programs evolved by linear variants of GP are represented as linear strings that are decoded and expressed as nonlinear entities (Oltean and Grossan 2003). A linear-GP system can run several orders of magnitude faster than comparable tree-based interpreting systems. The enhanced speed of the linear variants of GP permits conducting many runs in realistic timeframes. This leads to deriving consistent and high-precision models with little customization. GEP is one of the robust linear-GP techniques proposed by Ferreira (2001). While the traditional GP representation is based on the evaluation of a single tree (model) expression, GEP evolves computer programs of different sizes and shapes encoded in linear chromosomes of fixed length. The evolved programs are then expressed as parse trees of different sizes and shapes. These trees are called GEP expression trees (ETs) (Alavi and Gandomi 2011; Alavi et al. 2016). The nature of GEP allows the evolution of more complex programs composed of several subprograms. Function set, terminal set, fitness function, control parameters, and termination condition are the main components of GEP. Each GEP gene contains a list of symbols with a fixed length that can be any element from a function and terminal set. Comprehensive descriptions of the GEP algorithm can be found in (Alavi and Gandomi 2011; Alavi et al. 2016).

1.4.2 GEP-Based Formulation of PCI

The available database for Missouri airports was used to develop the GEP prediction models. Figure 1.8 shows the variation of PCI with respect to smartphone-measured IRI on the centerline lanes of the airfield pavements at the speed of 30 mph. For the GEP analysis, the database was randomly divided into the training and testing data. In the present study, 19 records were used for the training process and the remaining 7 datasets were taken as the testing data. The GEP parameters were varied across different runs to arrive at an optimal solution. Various parameters used in the present GEP algorithm are given in Table 1.4. The parameter setting was based on some previously suggested values (Alavi and Gandomi 2011; Alavi et al. 2016) and also after making several

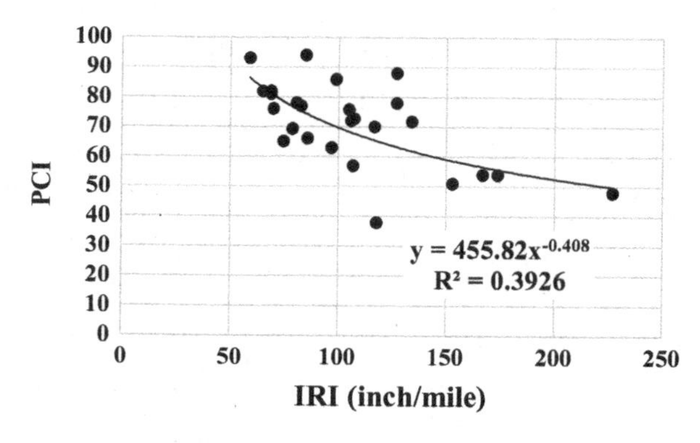

Figure 1.8 Variation of PCI with respect to smartphone-measured IRI.

Table 1.4: Parameter Settings for the GEP Algorithm

Parameter Settings		
General	Chromosomes	20, 40, 80
	Genes	2, 4, 6
	Head Size	6, 8, 10
	Linking Function	Addition
	Function set	+, −, ×, /, √, power, exp, sin, cos
Complexity Increase	Generations without Change	2000
	Number of Tries	3
	Max. Complexity	5
Genetic Operators	Mutation Rate	0.00138
	Inversion Rate	0.00546

preliminary runs and observing system performance. The number of programs in the population is set by the number of chromosomes (population size). The chromosome architectures of the models evolved by GEP include head size and number of genes. The head contains symbols representing both functions and terminals. The head size determines the complexity of each term in the evolved model. In other words, this parameter determines the upper limit for the size of the programs encoded in the gene. The number of terms in the model is determined by the number of genes per chromosome. Each gene codes for a different sub-expression tree or sub-ET. Different optimal levels were considered for the head size and number of genes parameters as tradeoffs between the running time and the complexity of the evolved solutions (Alavi and Gandomi 2011; Alavi et al. 2016). For the number of genes greater than one, the addition linking function was used to link the mathematical terms encoded in each gene. Referring to Table 1.4, there are $3 \times 3 \times 3 = 27$ different combinations of the algorithm parameters. A minimum of three replications were carried out for each parameter combination. The period of time acceptable for evolution to occur without improvement in best fitness was set via the generations without change parameter. After 2000 generations considered herein, a mass extinction or a neutral gene was automatically added to the model. The program was run until there was no longer significant improvement in the performance of the models.

The best GEP model for predicting PCI in term of the smartphone IRI is as follows:

$$\begin{aligned} \mathrm{PCI} &= 85.485 + \cos\left(-6.02 + 8.1679 IRI^2\right) \\ &+ 8.756\sin\left(IRI\right)^3 + 5.618\cos\left(IRI^3\right) - 4.922\cos\left(\left(IRI + 6.036407\right)^3\right) \\ &- \frac{SIN\left(IRI\right) * 11.178 + IRI - 5.056}{-5.938} \end{aligned} \tag{1.1}$$

As seen, the developed model is a highly nonlinear equation. It was generated by the GEP algorithm after controlling millions of linear and nonlinear models. Thus, it can efficiently consider the interactions between the IRI and PCI. Note that excluding trigonometric functions during the GEP analysis resulted in remarkably lower performance. Figure 1.9 shows the predictions made by the proposed model on the training and testing data. Based on a logical hypothesis (Smith 1986), if a model gives $|R| > 0.8$, and the error value (e.g., MAE and RMSE) is at the minimum, there is a strong correlation between the predicted and measured values. As seen in Figure 1.9, the GEP model has an acceptable

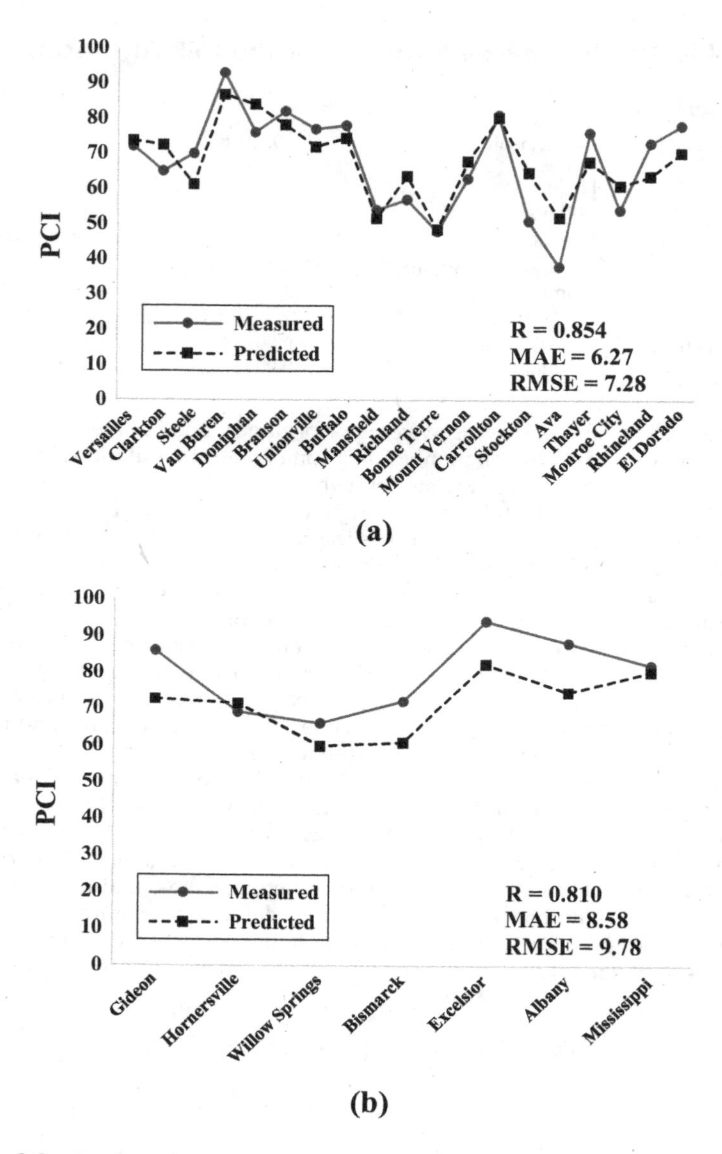

Figure 1.9 Predicted PCI using the GP model. (a) Training data. (b) Testing data.

performance with R values equal to 0.854 and 0.810 for the training and testing data, respectively. As more data become available, including those for other airports, the proposed model can be improved to make more accurate predictions over a wider range.

1.5 CONCLUSIONS

This project presents a new approach for the estimation of pavement roughness using acceleration data recorded by an Android-based smartphone application.

A calibration study was first performed on a test road near Excelsior Springs, MO. After the calibration study was completed, the proposed technology was implemented to determine the IRI values at 26 Missouri state funded general aviation airports. The IRI values were reported for the right, centerline, and left lanes of the airfield pavements. A wide range of IRI values was observed in the airfield pavements. Bonne Terre Municipal Airport pavement was found to be the least smooth segment with an average IRI of 227 inches/mile. while Bollinger-Crass Memorial Airport pavement had the best smoothness with an average IRI of 59 inches/mile. It can be reasonably concluded from this research that the proposed smartphone application has the potential to be an effective low-cost tool for assessing airport pavement condition. The smartphone estimated IRI values were close to those measured by the sophisticated ARAN system. The obtained trends agreed well with the construction and maintenance records of the airports. A GP-based model was developed to predict PCI based on the IRI values measured from the smartphone application. Based on the obtained trends, there is a reasonable correlation between the smartphone-based IRI roughness results and PCI.

For implementation, it is suggested that an improved graphical user interface (GUI) be developed for the smartphone roughness capture app, geared toward ease-of-use for airport managers. The study results suggest several ways in which this research can be enhanced in the future. In this study, only one smartphone model (Samsung Galaxy S8), one type of smartphone car mount, and one vehicle type (SUV) were used for data collection to reduce the uncertainties. A more robust approach could be developed by including a large number of smartphones and a fleet of vehicles to collect pavement roughness data through crowd sourcing. Moreover, estimating IRI based on the aircraft cab acceleration data may lead to more realistic results. Finding a sound correlation between the smartphone-based IRI, PCI, and Boeing bump index (BBI) could be an interesting topic for future research.

ACKNOWLEDGMENTS

The authors would like to thank the MoDOT for sponsoring this research. The authors express their gratitude to Andy Hanks, Jen Harper, and others at MoDOT for their assistance with the records review, coordination with the airports, and project guidance. The authors also thank Henry Brown, Henry Sills, Amanda Mesa, and Elizabeth Okenfuss at the University of Missouri, and Brian Aho, Jay Bledsoe, Bill Vavrik, Norman Adair, and others at ARA for collaborating on this project. Finally, the authors appreciate the assistance of the airport managers in providing information for the records review and helping to coordinate the field inspections.

REFERENCES

Alavi, A. H., A.H. Gandomi, A robust data mining approach for formulation of geotechnical engineering systems. *Eng. Comput.*, 2011, 28 (3):242–274.

ACI-NA, America's airports are vital economic hubs. Airports council international – North America (ACI-NA), white paper. 2010. [Online] 2011. https://aci-na.org/newsroom/press-releases/new-study-america%E2%80%99s-airports-are-vital-economic-hubs.

Adcock A. D., W. C. Dass, J. W. Rish III, Ground-penetrating radar for airfield pavement evaluations. In: T. M. Cordell, R. D. Rempt (Eds): *Nondestructive Evaluation of Aging Aircraft, Airports, Aerospace Hardware and Materials*. In: *Proceeding of SPIE*, 1995, 2455:373–384.

Aisopoulos, Paul J., Suspension system. [Online] 2011. www.ip-zev.gr/files/teaching/T3-5_Suspension_system.pdf.

Alavi A.H., H. Hasni, N. Lajnef, K. Chatti, F. Faridazar, Damage detection using self-powered wireless sensor data: An evolutionary approach. *Measurement*, 2016, 82:254–283.

Arraigada, M., M.N., Partl S.M. Angelone, F. Martinez, Evaluation of accelerometers to determine pavement deflections under traffic loads. *Materials and Structures*, 2009, 42:779–790.

ASTM D5340-12, American society for testing and materials (ASTM) standard D5340 – 12 "Standard Test method for Airport Pavement Condition Index Surveys," U.S.A., 2012.

Banzhaf W., P. Nordin, R. Keller, F. Francone, *Genetic Programming: An Introduction on the Automatic Evolution of Computer Programs and its Application*, Morgan Kaufmann, 1998.

Butt A., V.H.P. Pemmaraju, Pavement performance, FAA PAVEAIR and runway life extension. Presented to 2014 FAA Worldwide Airport Technology Transfer Conference, Galloway, NJ, U.S.A., August 5–7 2014.

Coello C.C.A., G.B. Lamont, D.A. Van Veldhuizen, Evolutionary algorithms for solving multi-objective problems. In: *Genetic and Evolutionary Computation*, 2nd ed., Springer, New York, 2007.

Douangphachanh V, H. Oneyama, Estimation of road roughness condition from smartphones under realistic settings. *Proceedings of the 13th International Conference on ITS Telecommunication (ITST)*, 2013, pp. 427–433.

Duan C., R. Wu, J. Liu, Estimation of airfield pavement void thickness using GPR. *Proceedings of the 3rd International Asia-Pacific Conference on Synthetic Aperture Radar (APSAR)*, Seoul, Korea, 2011, pp. 1–4.

Ferreira C., Gene expression programming: A new adaptive algorithm for solving problems, *Complex Syst.*, 2001, 13 :87–129.

Fogel L.J., A.J. Owens, M.J. Walsh, *Artificial Intelligence Through Simulated Evolution*, Wiley, New York, 1996.

Google. Google Maps. www.google.com/maps.

Hajek J., J.W. Hall, D.K. Hein, Common airport pavement, maintenance practices: A synthesis of airport practice. *ACRP Synthesis 22, Airport Cooperative Research Program (ACRP)*, Transportation Research Board, Washington, DC, 2011.

Hall J., R. Speir, H. Shirazi, E. Mustafa, I. Song, Performance trends in airport runway pavements. *Proceedings FAA Worldwide Airport Technology Transfer Conference*, Galloway, NJ, U.S.A., August 5–7 2014.

Islam, M.S. Development of a smartphone application to measure pavement roughness and to identify surface irregularities. Ph.D. Dissertation, University of Illinois at Urbana-Champaign, 2015.

Islam S., W. Buttlar, Effect of pavement roughness on user costs, *Transp. Res. Rec.*, 2012, 2285:47–55.

Islam, M.S., W.G. Buttlar, R.G. Aldunate, W.R. Vavrik, Improvements in pavement roughness measurement using smartphone application. *J. Trans. Res. Board*, 2016, 16-6729.

Islam, M.S., W.G. Buttlar, R.G. Aldunate, W.R. Vavrik, Measurement of pavement roughness using an android-based smartphone application. *J. Trans. Res. Board*, 2016, 2457:30–38.

Koza J.R., *Genetic Programming, on the Programming of Computers by Means of Natural Selection*, MIT Press, Cambridge, MA, 1992.

Mitchell T., Does machine learning really work? *AI Mag*, 1997, 18 (3):11–20.

Momeni E., R. Nazir, D. Jahed Armaghani, H. Maizir. Prediction of pile bearing capacity using a hybrid genetic algorithm-based ANN. *Measurement*, 2014, 57:122–131.

Oltean M., C. Grossan, A comparison of several linear genetic programming techniques, *Adv. Complex Syst.*, 2003, 14 (4):1–29.

Ozer E., Multisensory smartphone applications in vibration-based structural health monitoring. Ph.D. Dissertation, Columbia University, 2016.

Parsons T., B.A. Pullen, Relationship between joint spacing and distresses present. *Proceedings 2014 FAA Worldwide Airport Technology Transfer Conference*, Galloway, NJ, USA, August 5–7 2014.

Patra J.C. An artificial neural network-based smart capacitive pressure sensor. *Measurement*, (1997), 22:113–121.

Rabaiotti, C., D. Hauswirth, F. Fischli, M. Facchini, A. Puzrin, Structural health monitoring of airfield pavements using distributed fiber optics sensing. *Proceedings of the International Conference on Smart Monitoring, Assessment and Rehabilitation of Civil Structures (SMAR 2017)*, Zurich, Switzerland, 2017.

Sayer, M.W., T.D. Gilespie, W.D.O. Paterson, *Guidelines for Conducting and Calibrating Road Roughness Measurements*, World Bank, Washington, DC: World Bank Technical Paper 46, 1986.

Schwefel H.P., Evolutions strategie und numerischeOptimierung, Dissertation, TechnischeUniversität Berlin, 1975.

Smith G.N., *Probability and Statistics in Civil Engineering*, Collins, London, 1986.

Stribling J., W. Buttlar, M.S. Islam, Use of smartphone to measure pavement roughness across multiple vehicle types at different speeds. *The 96th TRB Annual Meeting*, Washington, DC, 2016.

U.S. Department of Transportation, 1999 Status of the nation's highways, bridges and transit: Conditions and performance. U.S. Department of Transportation, Federal Highway Administration, Washington, DC, 2000.

Unal M., M. Onat, M. Demetgul, H. Kucuk, Fault diagnosis of rolling bearings using a genetic algorithm optimized neural network. *Measurement*, 2014, 58:187–196.

Weise T., Global optimization algorithms – theory and application. Germany: it-weise.de (self-published), 2009. [Online]. http://www.it-weise.de.

White G., New airport pavement technologies from the USA. *Proceedings 2014 Australian Airport Association National Conference*, Gold Coast, Australia, 23–27 November, Australian Airports Association, 2014, pp. 23–27.

Yang S., H. Ceylan, K. Gopalakrishnan, S. Kim, Smart airport pavement instrumentation and health monitoring. *Proceedings 2014 FAA Worldwide Airport Technology Transfer Conference*, Galloway, NJ, U.S.A., August 5–7 2014.

Zang C., M. Imregun, Structural damage detection using artificial neural networks and measured FRF data reduced via principal component projection, *J. Sound Vib.*, 2001, 242 (5):813–827.

2 Global Satellite Observations for Smart Cities

Zhong Liu, Menglin S. Jin, Jacqueline Liu, Angela Li, William Teng, Bruce Vollmer, and David Meyer

CONTENTS

2.1 INTRODUCTION

Over the past several decades, the number of megacities (exceeding 10 million people in population) has been rapidly growing around the world as a result of rapid economic growth and unprecedented urbanization (United Nations 2014). For example, in Asia alone, more than 30 cities (Figure 2.1) are listed as megacities (e.g., Tokyo, Shanghai, Guangzhou), demanding effective management for city planning, operations, and disaster mitigation. The smart city approach requires data and information to be collected from multiple sources and to be integrated with modern technologies, providing a new and cost-effective way for decision-makers to manage cities in different sizes around the world as well as making information publicly available for city residents.

Environmental information at different spatial and temporal scales (e.g., ranging from local to regional and from real-time to climate) is one of critical sources for city planning, management, and disaster mitigation (Seto 2011). Each year, severe weather events (e.g., heavy rain or snowfall, tropical cyclones, heat waves)

Figure 2.1 Cities in Asia viewed from the 2016 annual NASA black marble, a nighttime view of the Earth, derived from a composite of data from the Visible Infrared Imaging Radiometer Suite (VIIRS) instrument on board the joint NASA/NOAA Suomi National Polar-orbiting Partnership (Suomi NPP) satellite (Lee et al. 2006). (Credit: NASA Worldview).

can strike a city around the world without warning and cause severe damage to a city's infrastructure as well as interrupt people's daily life. Effective management of water, air pollution, energy, etc. requires environmental data to be available at any time and on demand. Nonetheless, making data available for timely and easy access is critically important for effective city planning and management activities (Seto 2011).

Climate change can have a profound impact on cities around the world. For example, for cities in tropical and sub-tropical regions, changes in heavy rainfall amount as well as tropical cyclone size, frequency, and intensity can impact a city's operations, infrastructure, development, and long-term planning. Sea level rise is another major concern for city managers and residents in coastal cities. To be able to monitor and predict such change is critical for a city's planning and operations and all cannot be done without environmental data.

On the other hand, studies show that cities, especially large cities, can have an impact on local weather (e.g., Zhang et al. 2017; Jin et al. 2005, 2010, 2011; Shepherd and Jin 2004; Seto and Shepherd 2009; Kauffmann et al. 2007; Guo et al. 2016). Activities from urbanization and city development can dramatically modify its surrounding natural environment and landscape. For example, Urban Heat Islands (UHI) provide manmade heat sources that can change its surrounding atmospheric environment and potentially fuel severe weather. Air pollution that is associated with automobiles, industries, etc. supplies the atmosphere with aerosols that could modify meteorological processes such as clouds and precipitation (Jin et al. 2005, 2011).

Nonetheless, the few examples above have shown the importance of environmental data in city's planning and management. Such data consist of multi-disciplines at multi-scales in space and time. Traditional ground-based observations (e.g., rain gauges, automatic weather stations) have limitations because it is costly and time-consuming to deploy such measurements, especially at regional and global scales. On the other hand, satellite-based measurements, in combination with ground-based measurements, can overcome many of these difficulties and provide environmental data at multiple-scales (Seto 2007, 2011; Boucher and Seto 2009).

As mentioned earlier, the smart city approach requires collection of interdisciplinary data and information from multiple sources and integration with modern technologies to provide a new and cost-effective way for researchers and decision-makers to study and manage cities. In this book chapter, we introduce NASA satellite-based global and regional observations with emphasis on the hydrologic cycle (e.g., precipitation, wind, temperature, soil moisture) for smart cities. These products, consisting of both near-real-time and historical datasets, are publicly available free of charge and can be used for global and regional research and applications. Examples of using these datasets in smart cities are included. The chapter is organized as follows, first, a brief overview of NASA global satellite-based data products, followed by data services and tools, two examples of using satellite-based datasets in megacities, and finally summary and future plans.

2.2 OVERVIEW OF NASA SATELLITE-BASED GLOBAL DATA PRODUCTS FOR SMART CITIES

Significant progress has been made in satellite Earth's observations since the first successful launch of weather satellite, the Television Infrared Observation Satellite (TIROS), by NASA on April 1, 1960. In particular, NASA's Earth Observing System (EOS) is a coordinated series of polar-orbiting and low inclination satellites for long-term global observations of the land surface, biosphere, solid Earth, atmosphere, and oceans to enable an improved understanding of the Earth as an integrated system (NASA 2017a). At present, there are ~28 active satellite missions currently in space to provide observations to scientific and application users around the world (NASA 2017a). The Earth Observing System Data and Information System (EOSDIS) currently hosts ~22 PB of Earth Observation (EO) data at 12 DAACs (Distributed Active Archive Centers) and it is expected to grow rapidly over the coming years, to more than 37 (246 PB) PB by 2020 (2025) (NASA 2017b). Such large EO data archive is an important asset for environmental research and applications around the world including smart cities because the smart city approach requires multi-disciplinary datasets collected from multiple sources.

NASA's EOS and other past NASA satellite mission data are available and distributed by the EOSDIS, with major facilities at twelve DAACs located throughout the United States (NASA 2017b). Scientific disciplines at twelve DAACs include atmosphere, cryosphere, human dimensions, land, ocean, calibrated radiance, solar radiance, etc. Table 2.1 lists the twelve DAACs and their discipline-oriented data archives (NASA 2017b). For example, ocean winds and sea surface temperature data are available at the Physical Oceanography DAAC in the Jet Propulsion Laboratory. Lighting data from the NASA-JAXA Tropical Rainfall Measuring Mission (TRMM) are archived at the Global Hydrology Resource Center (GHRC) DAAC. As of this writing, a large collection of datasets and services are available at the twelve NASA data centers. However, due to page limit, it is difficult to give a detailed description for each DAAC. Here we

Table 2.1: EOSDIS DAACs and Their Archived Products (NASA 2017b)

Distributed Active Archive Center (DAAC)	Data Products
Alaska Satellite Facility (ASF) DAAC	SAR products, sea ice, polar processes, geophysics, etc.
Atmospheric Science Data Center (ASDC)	Radiation budget, clouds, aerosols, tropospheric chemistry, etc.
Crustal Dynamics Data Information System (CDDIS)	Space geodesy, solid Earth, etc.
Global Hydrology Resource Center (GHRC) DAAC	Hydrologic cycle, severe weather interactions, lightning, atmospheric convection, etc.
Goddard Earth Sciences Data and Information Services Center (GES DISC)	Global precipitation, solar irradiance, atmospheric composition and dynamics, global modeling, etc.
Land Processes DAAC (LP DAAC)	Surface reflectance, land cover, vegetation indices, etc.
Level 1 and Atmosphere Archive and Distribution System (LAADS) DAAC	MODIS and VIIRS Level-1 and atmosphere data products
National Snow and Ice Data Center (NSIDC) DAAC	Snow and ice, cryosphere, climate interactions, sea ice, etc.
Oak Ridge National Laboratory (ORNL) DAAC	Biogeochemical dynamics, ecological data, environmental processes, etc.
Ocean Biology DAAC (OB.DAAC)	Ocean biology, sea surface temperature, etc.
Physical Oceanography DAAC (PO. DAAC)	Gravity, sea surface temperature, ocean winds, topography, circulation, and currents, etc.
Socioeconomic Data and Applications Data Center (SEDAC)	Human interactions, land use, environmental sustainability, geospatial data, etc.

focus on the Goddard Earth Sciences and Data and Information Services Center (GES DISC), located in Greenbelt, Maryland, because it archives a large amount of interdisciplinary datasets in comparison to other DAACs. Datasets used in the examples in this article are archived at the GES DISC.

2.2.1 Satellite-Based Data Products at the GES DISC

The GES DISC hosts global and regional satellite-based interdisciplinary data products from these scientific disciplines: precipitation, solar irradiance, atmospheric composition and dynamics, global modeling, etc. Currently, over 2700 unique data products are archived at the GES DISC and distributed to the public. Given such a large collection of products and space limitation, we can only present a brief overview of several major projects that are closely related to research and applications for smart cities.

2.2.1.1 Multi-Satellite and Multi-Sensor Merged Global Precipitation Products

Over the years, algorithms that utilize multi-satellites and multi-sensors (i.e., microwave and geostationary infrared (IR) sensors), or blended methods, have been developed to overcome a very limited spatial and temporal coverage from any single satellite (Adler et al. 2003; Huffman et al. 2007, 2009, 2010, Huffman and Bolvin 2012, 2013; Joyce et al. 2004; Mahrooghy et al. 2012; Hong et al. 2007, Sorooshian et al. 2000; Behrangi et al. 2009; Aonashi et al. 2009) and products are widely used in hydrometeorological research and applications. For example,

Table 2.2: Global Gridded Multi-Sensor and Multi-Satellite Precipitation Products (Liu et al. 2012, 2017)

Dataset	Description	Date Range	Spatial Resolution and Coverage	Temporal Resolution
The Integrated Multi-satellitE Retrievals for GPM (IMERG) – "Early, Late, and Final"	Rain rate from multi-satellite, multi-sensor, and gauge measurements	1998-01-01 - present	Gridded 10 km, global, initially 60°N-60°S	Half-hourly, daily, and monthly
TMPA (Near-real-time, Research)	Rain rate from multi-satellite and multi-sensor measurements	1998-01-01 - present	Gridded 25 km, 60°N-60°S (Research: 50°N-50°S)	3-hourly, daily, and monthly

the TRMM Multi-Satellite Precipitation Analysis (TMPA) products in Table 2.2 (Huffman et al. 2007, 2010; Huffman and Bolvin 2012, 2013), developed by the Mesoscale Atmospheric Processes Laboratory at NASA Goddard Space Flight Center, provide precipitation estimates at 3-hourly and monthly temporal resolutions on a 0.25-degree × 0.25-degree grid available from January 1998 to present. The TMPA consists of two products: near-real-time (3B42RT, spatial coverage: 60°N–60°S) and research-grade (3B42, spatial coverage: 50°N–50°S). The former is less accurate, but provides quick precipitation estimates suitable for near-real-time monitoring and modeling activities (e.g., Wu et al. 2012). The latter, available approximately two months after observation, is calibrated with gauge data, different sensor calibration, and additional post-processing in the algorithm. The resulting product is more accurate and suitable for research (Huffman et al. 2007, 2010). Over the years, the TMPA products have been widely used in various research and applications (e.g., Wu et al. 2012; Bitew et al. 2012; Gourley et al. 2011; Su et al. 2011; Gianotti et al. 2012; Tekeli 2017; Engel et al. 2017; Tan and Duan 2017).

During the GPM era, the Integrated Multi-satellitE Retrievals for GPM (IMERG) product suite (Huffman et al. 2017) not only addresses limited spatial and temporal coverage issues in TRMM but also has been significantly improved comparing the TMPA products in terms of spatial and temporal resolutions, i.e., from 0.25 degree to 0.1 degree and from 3-hourly to half-hourly (Table 2.2). The IMERG suite contains of three output products, "Early satellites" (lag time: ~4 hours), "Late satellites" (lag time: ~12 hours), and the final "Satellite-gauge" (lag time: ~2.5 months) along with additional new input and intermediate files, creating a new opportunity for research and applications. The retro-processing of the IMERG product suite back to the TRMM era will be released in the near future.

2.2.1.2 Global and Regional Land Data Assimilation Products

Global and regional Land Data Assimilation System (LDAS) data products include optimal fields of land-surface states and fluxes (NASA 2017c). The fields are generated by ingesting satellite- and ground-based observational data products and using advanced land-surface modeling and data assimilation techniques (NASA 2017c). A methodology is to implement a LDAS that consists of land-surface models (uncoupled from an atmospheric model). These land-surface models are forced with observations and thus the model results are not affected by Numerical Weather Prediction forcing biases (NASA 2017c). This research has been implemented using existing Surface Vegetation Atmosphere

Transfer Schemes by NOAA, NASA/GSFC, NCAR, Princeton University, and the University of Washington at 1/8th-degree resolution across central North America and at 1/4th-degree resolution globally to evaluate these critical science questions (NASA 2017c). These LDAS systems have been run retrospectively starting in January 1979 and continue in near real-time, and are forced with gauge precipitation observations, satellite precipitation data, radar precipitation measurements, and output from Numerical Weather Prediction models (NASA 2017c). Model parameters are derived from the existing high-resolution vegetation and soil coverages. The LDAS model results support water resources applications, numerical weather prediction studies, numerous water and energy cycle investigations, and also serves as a foundation for interpreting satellite and ground-based observations (NASA 2017c). Eventually, in situ or remotely-sensed observations (soil moisture, temperature, snow) of LDAS storages and fluxes (including evaporation, sensible heat flux, runoff) will be used to further validate and constrain the LDAS predictions using data assimilation techniques (NASA 2017c). The GES DISC hosts the archive of data products from GLDAS, NLDAS, NCA-LDAS, and FLDAS (NASA 2017c).

2.2.1.3 Modern-Era Retrospective Analysis for Research and Applications (MERRA) Products

The Modern-Era Retrospective analysis for Research and Applications, Version 2 (MERRA-2), has been developed at the NASA Global Modeling and Assimilation Office (GMAO) at the NASA Goddard Space Flight Center (NASA 2017d). MERRA-2 provides global data beginning in 1980 and runs a few weeks behind real-time (NASA 2017d). Alongside the meteorological data assimilation using a modern satellite database, MERRA-2 includes an interactive analysis of aerosols that feed back into the circulation, uses NASA's observations of stratospheric ozone and temperature (when available), and takes steps toward representing cryogenic processes (NASA 2017d). The MERRA project focuses on historical analyses of the hydrological cycle on a broad range of weather and climate time scales and places the NASA EOS suite of observations in a climate context (NASA 2017d). Compared to the previous version, advances have been made in the assimilation system that enables assimilation of modern hyperspectral radiance and microwave observations, along with GPS-Radio Occultation datasets (Riencker et al. 2011; Reichle and Liu 2014; Suarez and Bacmeister 2015; Takacs et al. 2015). MERRA-2 also includes advances in both the Goddard EOS Model, Version 5 (GEOS-5) and the Gridpoint Statistical Interpolation assimilation system and NASA ozone observations after 2005. MERRA-2 begins from 1980 to present.

There are two types of precipitation parameters in MERRA-2: a) precipitation from the atmospheric model (variable PRECTOT in the MERRA-2 data collection) and b) observation-corrected precipitation (variable PRECTOTCORR; Reichle and Liu 2014; Bosilovich et al. 2015). Observational data are introduced in the latter parameter due to considerable errors that propagate into land-surface hydrological fields and beyond (Reichle et al. 2011).

Bosilovich et al. (2015) have conducted a general evaluation of MERRA-2 precipitation estimates, including precipitation climatology, inter-annual variability, diurnal cycle, Madden–Julian Oscillation (MJO) events, global water cycle and U.S. summertime variability. Major findings (Bosilovich et al. 2015) are, 1) an overestimate of the modeled precipitation in the tropical west Pacific Ocean, the eastern tropical ITCZ (the Intertropical Convergence Zone) and the South Pacific convergence zone in DJF and JJA (Bosilovich et al. 2011, 2015); 2) Extreme values of modeled precipitation in the vicinity of high topography in the tropics;

3) An upward trend in the MERRA-2 time series exists and by contrast no trend is observed in GPCP (the Global Precipitation Climatology Project); 4) Larger modeled Precipitation Diurnal Cycle (PDC) amplitude is found over the high mountains; 5) The phases of modeled PDC are not well reproduced in several regions such as the United States Great Plains; 6) MJO signal from modeled precipitation is stronger than GPCP; 7) MERRA-2 can reproduce the observed precipitation and anomalies in U.S. summertime, reasonably well. Although the preliminary evaluation provides a basic understanding of the MERRA-2 precipitation products, evaluation for extreme rainfall events is missing and as a result, it is not clear about MERRA-2 precipitation behavior and characteristics in extreme events.

The complete list of MERRA-2 products along with documentation and more is available on the official MERRA-2 Web site (NASA 2017d). To facilitate data access, the GES DISC has developed several data services and tools to be described in the next section.

2.3 DATA SERVICES

Although a large collection of NASA global satellite data is available for research and applications around the world, many researchers find it challenging to discover, access and use NASA satellite remote sensing data (Liu and Acker 2017). Heterogeneous data formats, complex data structures, large-volume data storage, special programming requirements, diverse analytical software options, and other factors often require a significant investment in time and resources, especially for novices (Liu and Acker 2017). Over the years, data services have been developed at NASA's EOSDIS DAACs to improve NASA data discovery and access. First, an EOSDIS Web search interface (NASA 2017e) has been developed and anyone, with a Web browser, can access NASA data products at twelve DAACs through this interface. Figure 2.2 is a screenshot of the EOSDIS Web search interface, showing a search box where users can type

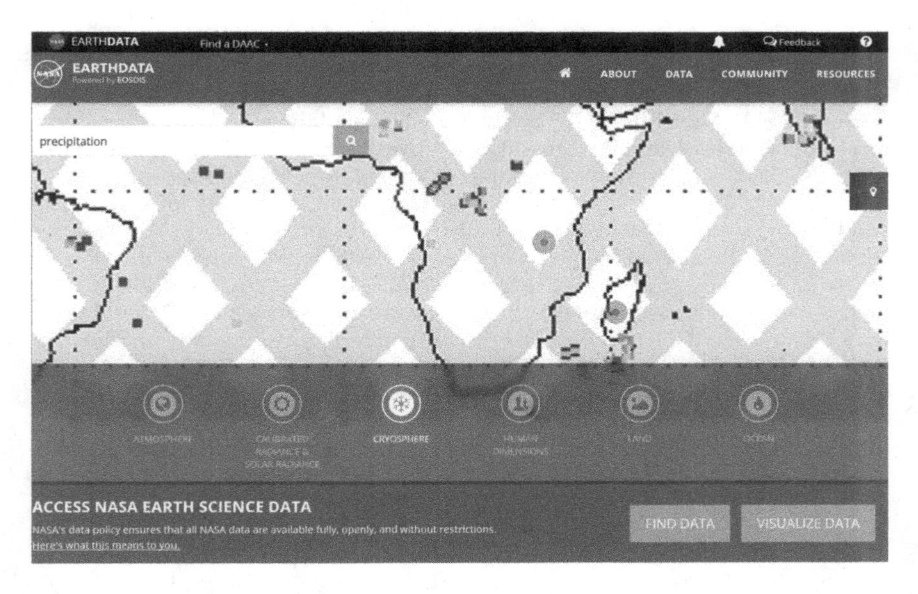

Figure 2.2 NASA Earthdata (NASA 2017b) provides a Web interface for searching and accessing NASA data at twelve EOSDIS DAACs.

in a data variable name such as precipitation. User registration is required for downloading data from all EOSDIS data services. Users can also visit each individual DAAC and use their Web interfaces to access discipline-oriented data products and services. Furthermore, special discipline-oriented data tools have been developed at DAACs and they are organized in the following categories: 1) Search and Order; 2) Data Handling; 3) Subsetting and Filtering; 4) Geolocation, Reprojection, and Mapping; and 5) Data Visualization and Analysis (NASA 2017f). EOSDIS DAACs also support Web services and various Web protocols for machine-to-machine data access and applications such as OPeNDAP (Open source Project for a Network Data Access Protocol), WMS (Web Map Service), GDS (GrADS Data Server), THREDDS, https, etc. To address data and science related issues and inquiries from users, DAACs provide user services including Frequent Asked Questions, data recipes, user forums, email or phone inquiry, etc. Due to space limitation, it is difficult to describe all data tools and services at the DAACs in one article. Since we focus on datasets for the hydrologic cycle in this article, data services at the GES DISC and other well-known services are presented.

2.3.1 Point-and-Click Online Tools

Over the years, surveys (e.g., Kearns 2017) and experience from user support services at the GES DISC show that non-expert users and those who occasionally use satellite-based products prefer point-and-click data tools in order to obtain graphic and data assessment results. As mentioned above, new dataset assessment activities may not be straight-forward and can be costly. Point-and-click tools provide fast and easy access to satellite-based data products for all users without the need of coding and downloading data and software. Here, we introduce two popular point-and-click online tools developed by NASA: the NASA Worldview and the NASA GES DISC Giovanni.

2.3.1.1 NASA's Worldview

NASA's Worldview is a tool developed by the NASA's EOSDIS project (Figure 2.3). It provides the capability to interactively browse global, full-resolution satellite imagery and then download the underlying data (NASA 2017g). Worldview contains 400+ NASA satellite-based products and most of them are updated within three hours of observation, essentially showing the entire Earth as it looks "right now" (NASA 2017g). Worldview is a user-friendly online tool to support time-critical application areas such as wildfire management, air quality measurements, and flood monitoring. Mobile access is also available for Worldview. Worldview is powered by the Global Imagery Browse Services to rapidly retrieve its imagery for an interactive browsing experience (NASA 2017g). Data in Worldview images can be exported as several popular image formats such as JPEG and PNG. Image data in GeoTIFF and KMZ are also available. Figure 2.3 is a screenshot of Worldview, showing the Web interface and Hurricane Harvey in Gulf of Mexico on August 24, 2017. Hurricane Harvey made landfall near Corpus Christ, Texas. A record-breaking flood in Houston, Texas was reported due to heavy rainfall from Hurricane Harvey's rainbands. Fatalities and enormous economic loss in Houston were reported as well.

2.3.1.2 NASA GES DISC Giovanni

Point-and-click tools can be further developed for in-depth data analysis and visualization. A new infrastructure system, the Geospatial Interactive Online Visualization and Analysis Infrastructure (Giovanni, NASA 2017h), has been

Figure 2.3 The NASA Worldview Web interface (NASA 2017g) showing Hurricane Harvey in Gulf of Mexico on August 24, 2017.

developed by the GES DISC to assist a wide range of users around the world with data access and evaluation, as well as with scientific exploration and discovery (Liu and Acker 2017; Acker and Leptoukh 2007). There are 8 disciplines and 74 measurements available in Giovanni and they are listed in Tables 2.3 and 2.4, respectively. Over 1,900 variables are currently available in Giovanni as of this writing and more are being added. Data variables in Giovanni are multi-missions and multi-disciplinary (Figure 2.4). Users can access these variables without downloading data and software (Liu and Acker 2017). Over the years, a wide range of activities of using Giovanni has been reported by users, ranging from classroom activities to scientific investigation. Over 1,700 peer-reviewed papers across various Earth science disciplines and other areas were published with help from Giovanni.

Giovanni has both Keyword and Faceted Search capabilities in its Web interface (Figure 2.4) for locating variables of interest. For example, a search for "precipitation" returns over 100 related variables (Figure 2.4). By using facets, one can filter for variables based on satellite missions (TRMM, GPM), instruments, spatial or temporal resolution, etc.

Table 2.3: Disciplines and Variables in Giovanni (NASA 2017h)

Disciplines (No. of Variables)
Aerosols (183)
Atmospheric Chemistry (79)
Atmospheric Dynamics (385)
Cryosphere (15)
Hydrology (997)
Ocean Biology (44)
Oceanography (48)
Water and Energy Cycle (1065)

Table 2.4: Measurements and Variables in Giovanni (NASA 2017h)

Measurement (No. of Variables)

- Aerosol Index (3)
- Aerosol Optical Depth (83)
- Air Pressure Anomaly (1)
- Air Pressure (51)
- Air Temperature (84)
- Albedo (21)
- Altitude (8)
- Angstrom Exponent (17)
- Atmospheric Moisture (114)
- Black Carbon (5)
- Buoyancy (2)
- CH_4 (16)
- CO (21)
- CO_2 (2)
- Canopy Water Storage (6)
- Chlorophyll (11)
- Cloud Fraction (32)
- Cloud Properties (75)
- Component Aerosol Optical Depth (7)
- Diffusivity (1)
- Dust (23)
- Emissivity (4)
- Energy (12)
- Erythemal UV (4)
- Evaporation Anomaly (1)
- Evaporation (44)
- Evapotranspiration (41)
- Flooding (3)
- Geopotential (11)
- Heat Flux (102)
- Height, Level (12)
- Incident Radiation Anomaly (2)
- Incident Radiation (70)
- Iron (2)
- Irradiance (6)
- Latent Heat Flux (5)
- Latent Heat (1)
- Mixed Layer Depth (2)
- NO_2 (2)
- Nitrate (2)
- OLR (19)
- Organic Carbon (8)
- Ozone (28)
- Particulate Matter (42)
- Phytoplankton (16)
- Precipitation Anomaly (2)
- Precipitation (107)
- Quality Info (1)
- Radiation, Net (56)
- Reflectivity (25)
- Runoff (63)
- SO_2 (4)
- SO_4 (4)
- Scattering Angle (4)
- Sea Salt (5)
- Sea Surface Temperature (3)
- Sensible Heat Flux (5)
- Sensible Heat (1)
- Snow/Ice (37)
- Soil Moisture (203)
- Soil Temperature (105)
- Statistics (24)
- Streamflow (1)
- Surface Runoff (1)
- Surface Temperature (55)
- Total AOD Climatology Anomaly (6)
- Total Aerosol Optical Depth (49)
- UV Exposure (1)
- Vegetation (9)
- Vorticity (2)
- Water Storage (1)
- Wind Stress Magnitude (4)
- Wind Velocity (7)
- Wind (72)

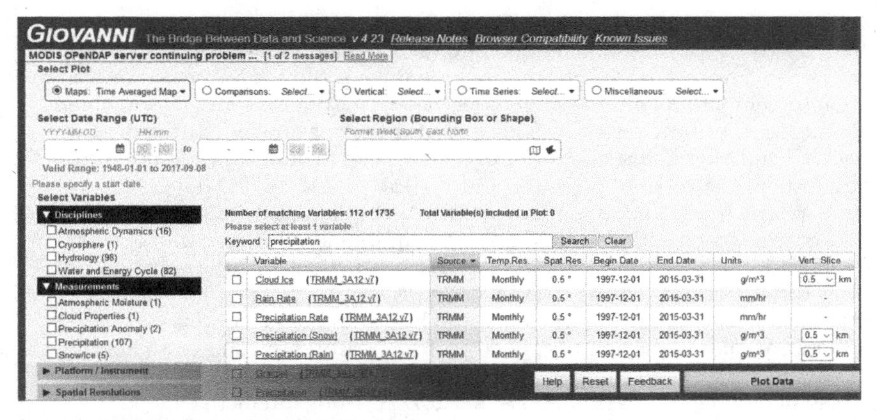

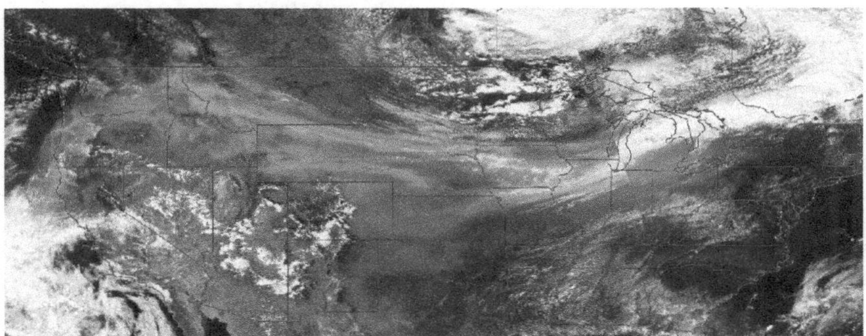

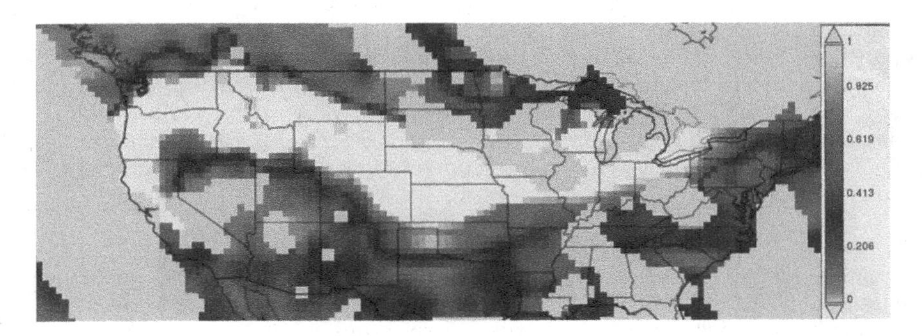

Figure 2.4 Top: The Web portal of NASA GES DISC Giovanni (NASA 2017h) with many features for easy locating a variable of interest, data analysis, and visualization; Middle: Corrected reflectance (true color) from MODIS Aqua on September 4, 2017, showing smoke from forest fires spreads across the United States and Canada (More detailed story is available at, https://earthobservatory.nasa.gov/IOTD/view.php?id=90899). Bottom: The combined dark target and deep blue aerosol optical depth (AOD) at 0.55 micron for land and ocean from MODIS Aqua on September 4, 2017. Combined with NASA Worldview, Giovanni provides analysis and visualization of the smoke aerosol for the event.

Many commonly used analytical and plotting capabilities (Liu and Acker 2017), used to capture spatial and temporal characteristics of datasets, are available in Giovanni. Mapping options include time-averaging, animation, accumulation (precipitation), time-averaged overlay of two datasets and user-defined climatology. For time series, options include of area-averaged, differences, seasonal and Hovmöller diagrams. Cross-sections include latitude-pressure, longitude-pressure, time-pressure, and vertical profile for 3-D datasets from Atmospheric Infrared Sounder and MERRA. For data comparison, Giovanni has built-in processing code for data sets that require measurement unit conversion and regridding. Commonly used comparison functions include map and time-series differences, as well as correlation maps and XY scatter plots (area-averaged or time-averaged). Zonal means and histogram distributions can also be plotted. Samples of the analytical and plotting features are shown in Figures 2.4 and 2.5.

Visualization features (Liu and Acker 2017) include interactive map area adjustment; animation; interactive scatter plots; adjustments of data range; change of color palette; contouring; and scaling (linear or log). The on-the-fly area adjustment feature allows an interactive and detailed examination of a result map without re-plotting data. Animations are helpful to track evolution of an event or seasonal changes. Interactive scatter plots allow identification of the geolocation of a point of interest in a scatter plot. Adjustments of any of these plots provide custom options to users.

To support increasing socioeconomic and GIS activities in Earth sciences, vector shapefiles have been added for countries, states in the United States, and major watersheds around the world. Available functions for shapefiles are time-averaged and accumulated maps, area-averaged time series, and histogram. Land-sea masks have been recently added.

All data files involved in Giovanni processing are listed and available in the lineage page. Available image formats are PNG, GeoTIFF and KMZ (Keyhole Markup Language) that can be used for different applications and software packages; for example, KMZ files can be conveniently imported into Google Earth where a rich collection of overlays is available. All input and output data are available in NetCDF, which can be handled by many off-the-shelf software packages. Furthermore, users can bookmark URLs generated by Giovanni processing for reference, documentation, or sharing with other colleagues.

2.3.2 Data Rod Services

Providing long-time series data to the hydrology community can be a challenge (Teng et al. 2016). In hydrology, Earth surface features are expressed as discrete spatial objects such as watersheds, river reaches, and point observation sites; and time varying data are contained in time series associated with these spatial objects. Long-time histories of data may be associated with a single point or feature in space. Most remote sensing precipitation products are expressed as continuous spatial fields, with data sequenced in time from one data file to the next. Hydrology tends to be narrow in space and deep in time, which poses a challenge during the GPM era. For example, to generate a one-year time series, one needs to pull all the 0.1 degrees, half-hourly IMERG product files, which can be time-consuming and not suitable for online data services due to the large volume of data.

The concept of data rods (Teng et al. 2016; Gallaher and Grant 2012; Rui et al. 2012, 2013) can be applied to this challenge. Teng et al. (2016) proposed two general solutions: 1) retrieve multiple time series for short time periods

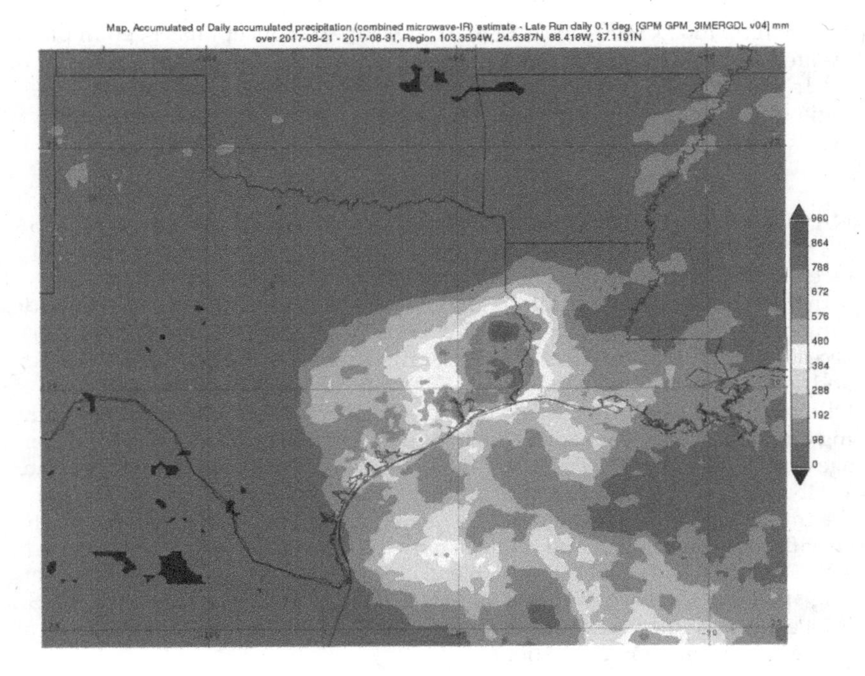

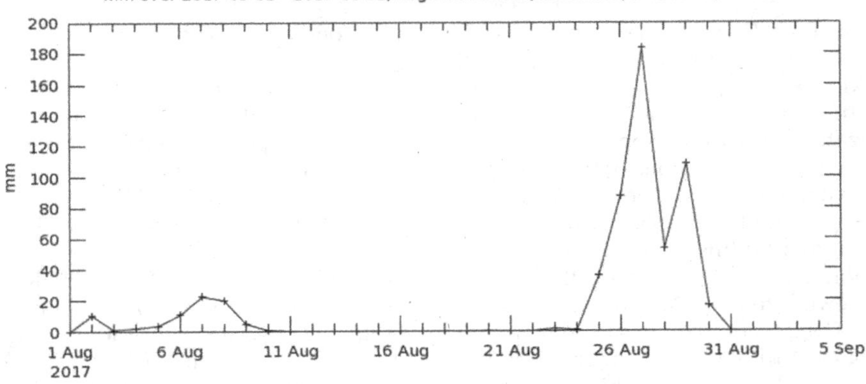

Figure 2.5 Sample graphics from Giovanni. Top: Accumulated rainfall (the GPM IMERG Late Run daily product) in mm from Hurricane Harvey (see Figure 2.3) in Houston, Texas between August 21–31, 2017. Bottom: Time series of area-averaged daily accumulated precipitation (mm/day) in Houston.

and stitch the multiple time series into desired single long-time series; and 2) reprocess (parameter and spatial subsetting) and archive data as a one-time cost approach. The resultant time series files would be geospatially searchable and could be optimally accessed and retrieved by any user at any time (Teng et al. 2016). One drawback for the data rod approach is that there are a lot of files to be generated and maintained. For example, for IMERG, the number of files will be 1300×3600=4,680,000. At present, the concept has been implemented in CUAHSI-HIS (Consortium of Universities for the Advancement of Hydrologic Science, Inc. Hydrologic Information System) and other hydrologic community tools (Rui et al. 2013) where TMPA data time series data can be accessed.

2.3.3 Other Web Data Services

NASA satellite-based data products at the GES DISC are also accessible (NASA 2017i) via other Web services and protocols including https (the data archive), OPeNDAP, WMS, GDS, etc. These protocols support for data downloading activities such as daily operations on the user's side. The https method provides direct access to product archives. OPeNDAP, WMS, GDS, etc. provide remote access to individual variables within datasets in a form usable by many tools and software packages such as IDV, McIDAS-V, Panoply, Ferret, GrADS, etc. OPeNDAP is a framework that simplifies all aspects of scientific data networking and makes local data accessible to remote locations regardless of local storage format (OPeNDAP 2017). OPeNDAP software is freely available to anyone. WMS is a standard Web protocol for serving georeferenced map images over the Internet generated by a map server using data from a GIS database and the specifications developed and published by the Open Geospatial Consortium in 1999 (WMS 2017). The GDS, is a stable, secure data server that provides subsetting and analysis services across the Internet (GDS 2017). The core of the GDS is OPeNDAP, a software framework used for data networking that makes local data accessible to remote locations.

2.4 EXAMPLES

2.4.1 The Pearl River Delta

The Pearl River Delta (PRD) is located in Guangdong province, the People's Republic of China (PRC). The PRD is a low-lying area surrounding the Pearl River estuary where the Pearl River flows into the South China Sea (Wikipedia 2017a). The region has experienced an economic boom and accelerated urbanization since the region was named as one of the three Special Economic Zones by the PRC and is one of the most densely urbanized regions in the world (Wikipedia 2017a). The city of Guangzhou has become a mega city, the 3rd largest in PRC and the largest in southern China, with a population of over 15 million. Adding the nearby cities, the total population in PRD is over 40 million, forming the so-called the Pearl River Delta Mega City. Figure 2.6 is the PRD viewed from the 2016 annual NASA black marble, a nighttime view of the Earth derived from a composite of data from the Visible Infrared Imaging Radiometer Suite (VIIRS) instrument on board the joint NASA/NOAA Suomi National Polar-orbiting Partnership (Suomi NPP) satellite. UHI effect is also quite visible from the MODIS-Terra monthly nighttime land-surface temperatures averaged between December 2016 and February 2017 (Figure 2.7). Giving economic activities at such scale, it is important to understand environment impacts that are associated with economic activities and EO data play an important role to provide such analysis and information. In this chapter, we present few examples regarding how NASA EO datasets are used in this region.

2.4.1.1 Typhoon Nida Rainfall

Formed on July 28, 2016, Typhoon Nida struck the Philippines in late July and made landfall in the PRD as a category-1 typhoon (Figure 2.8) in early August (Wikipedia 2017b). According to news reports, Nida caused heavy economic loss

Figure 2.6 Similar to Figure 2.1, except for the Pearl River Delta.

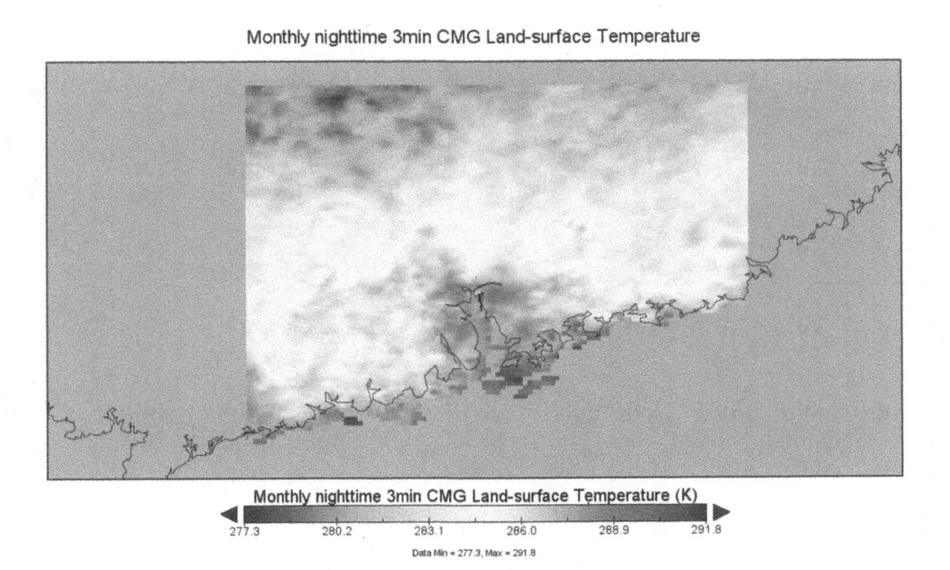

Figure 2.7 Map of MODIS-Terra monthly nighttime land-surface temperatures averaged between December 2016 and February 2017, showing visible light produced from anthropogenic sources (e.g., city lights) (Lee et al. 2006).

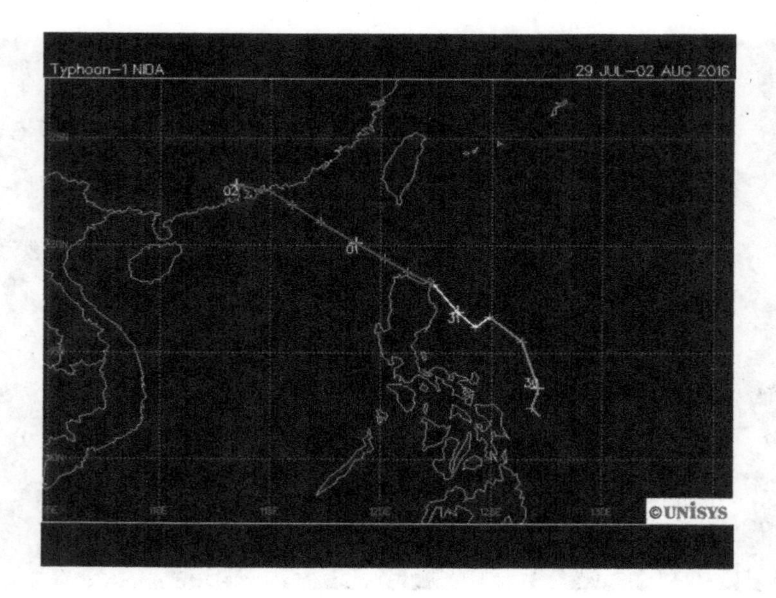

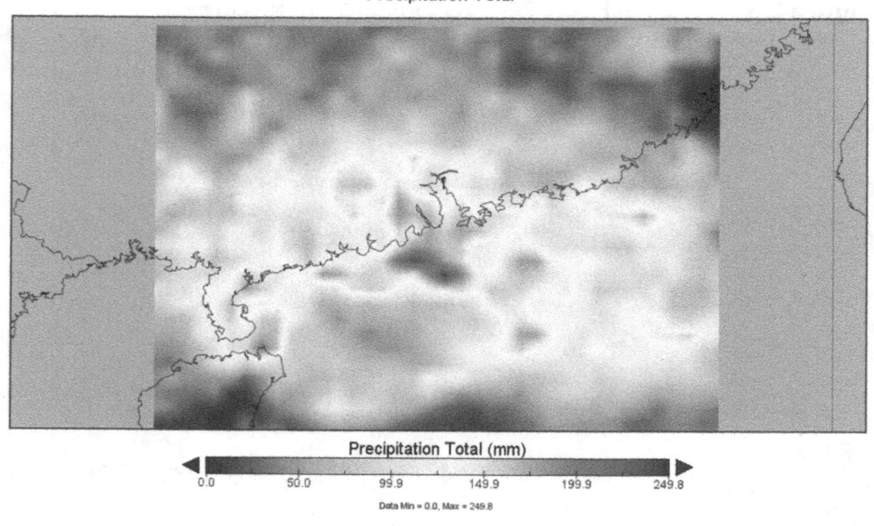

Figure 2.8 Top: Track of Typhoon Nida (source: Unisys Weather); Bottom: Rainfall total received from Typhoon Nida between August 1–5, 2016 (Rainfall product: the 3-hourly TMPA research product).

in the affecting countries and regions including the Philippines, the mainland of PRC, Hong Kong, and Vietnam (Wikipedia 2017b). In PRC alone, ~495,000 people in five southern provinces were affected. 37,000 people required evacuation and 2,100 needed emergency assistances (Wikipedia 2017b). Homes and crops were destroyed or damaged. People's lives in the region were severely disrupted, according to news reports (Wikipedia 2017b).

NASA EO data are available for Nida. At the GES DISC, precipitation from TMPA, GPM, IMERG, MERRA, GLDAS, etc. with different spatial and temporal resolutions are available for research and analysis. Figure 2.8 is an example of total rainfall in mm accumulated between August 1–5, 2016. It is seen that heavy rainfall was received in the western part of the PRD and the heaviest rainfall area is found in the adjacent ocean off the coast of the PRD (Figure 2.8). MERRA provides data such as wind, temperature, pressure, etc. for meteorological analysis. GLDAS provides data for hydrological research and analysis. Many of these datasets are available in the GES DISC Giovanni and ready for assessment, analysis, and visualization without the need to download data and software.

2.4.1.2 Atmospheric Composition Preliminary Analysis

As mentioned, the PRD has experienced an unprecedented economic growth and urbanization since 1979. The environmental conditions in the region have also experienced the changes, which requires EO data for research and analysis. MERRA uses various NASA satellite observations and the chemistry model to generate reanalysis products for research and applications. Figures 2.9 and 2.10 are the time series plots of sample MERRA-2 atmospheric chemistry variables for the PRD: monthly aerosol optical depth, SO2 surface mass concentration, CO surface concentration, and black carbon surface mass concentration, showing seasonal and inter-annual variations in the PRD over the past 30+ years. In Figure 2.9, it is seen that aerosol optical depth experienced a large increasing trend since 2000, followed by a decreasing trend after mid-2000 or so. According to reports (Wikipedia 2017c), the PRD region's GDP in 2005 was ~US$221.2 billion, compared to US$89 billion in 2000. Further investigation is needed to better understand and link these observation results with the economic growth in the region. The SO2 surface mass concentration experienced a steady increase before 2000, followed a rapid climb shortly after 2000 (Figure 2.9). The increasing period ended near 2010 and the concentration still remained high with fluctuations afterward (Figure 2.9). The similar trends are also found in CO surface concentration (Figure 2.10). The black carbon surface mass concentration (Figure 2.10) is quite similar to the aerosol optical depth in Figure 2.9. It is necessary to point out that these time series plots in Figures 2.9 and 2.10 are preliminary and need to be verified independently with ground measurements to ensure that biases and other issues are addressed properly.

2.4.2 Estimation of Hurricane Contribution to Annual Precipitation in Maryland

The state of Maryland, is located in the Mid-Atlantic region near the nation's capital of Washington DC. According the 2016 report (Censor Bureau 2016) from the United States Censor Bureau, the population of the region that consists of Washington–Arlington–Alexandria and DC–Virginia–Maryland–West Virginia, reached over 6 million in 2015 and became the 6th most populous metro area in the United States.

Despite the fact that major hurricanes (category 3 or above) rarely make landfall in the state, Maryland is indirectly influenced by hurricane remnants (NOAA 2017a) such as rainfall. Giving the global warming scenario, it is important to understand changes of hurricane size, track and intensity since all of them could have significant impacts on precipitation in Maryland. In this study, the TMPA precipitation products were used to assess hurricane precipitation. Unlike many land-only precipitation products, the TMPA products not only provide precipitation information over land but also over oceans, therefore, they

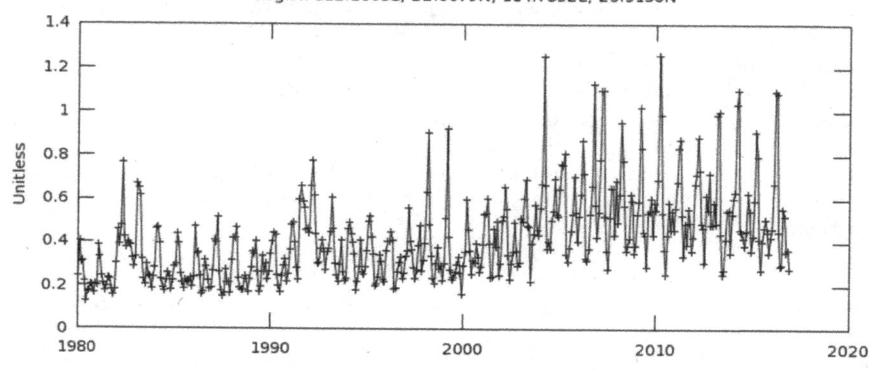

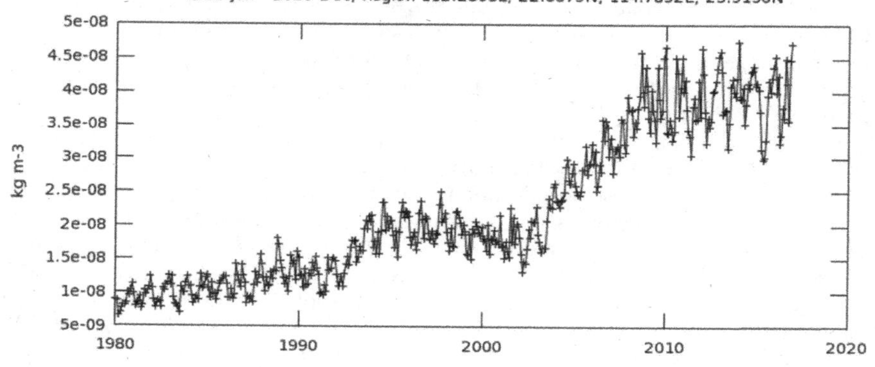

Figure 2.9 Time series from MERRA-2 showing seasonal and internal variations of monthly aerosol optical depth (top) and SO2 surface mass concentration (bottom) in PRD.

are suitable for this type of study. The objectives of this study are to estimate hurricane contribution to annual precipitation in Maryland and its inter-annual variation (Liu and Liu 2015). The methodology could be applied to other states or regions as well.

2.4.2.1 Data and Methods

The Version-7 3-hourly (3B42) and monthly (3B43) TMPA products are used in this study. 3B42 and 3B43 provide a near-global (50° N-S) coverage of both land and oceans and allow tracking of hurricane precipitation since some hurricanes can pass by Maryland over ocean without making landfall and influence the state with rain, winds, waves, etc.

The TMPA algorithm (Huffman 1997; Huffman et al. 2007, 2010; Huffman and Bolvin 2014) consists of multiple independent precipitation estimates from the TRMM Microwave Imager, Advanced Microwave Scanning Radiometer for EOSs, Special Sensor Microwave Imager, Special Sensor Microwave

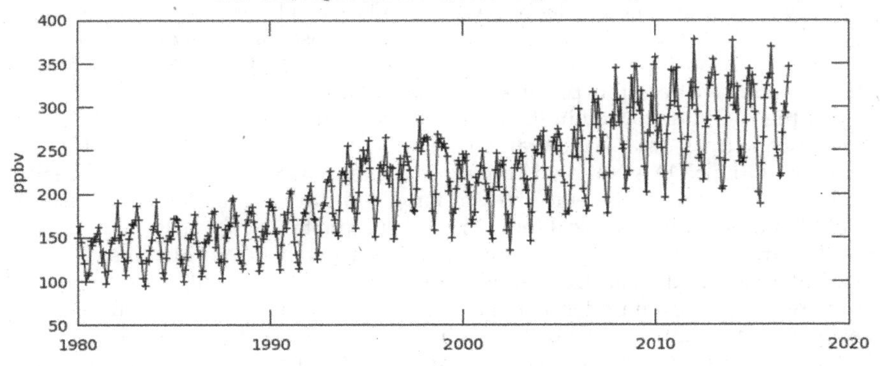

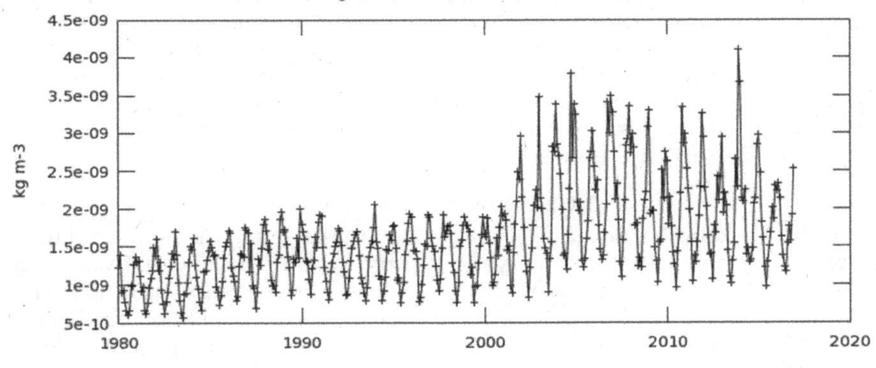

Figure 2.10 Similar to Figure 2.9, except for CO surface concentration (top) and black carbon surface mass concentration (bottom).

Imager/Sounder, Advanced Microwave Sounding Unit, Microwave Humidity Sounder, Microwave-Adjusted Merged Geo-Infrared, and monthly accumulated rain gauge analysis from the Global Precipitation Climate Centre (GPCC). The preprocessing (Huffman and Bolvin 2014) of the TMPA products is as follows: (a) all input passive microwave (PMW) products mentioned above are inter-calibrated to TRMM Combined Instrument precipitation estimates (TRMM product 3B31); (b) the IR estimates are computed using monthly matched microwave-IR histogram matching; (c) then missing data in individual 3-hourly merged-microwave fields are filled with the IR estimates. When the preprocessing is complete, the 3-hourly multi-satellite fields are summed for the month and combined with the monthly GPCC gauge analysis using inverse-error-variance weighting to form the best-estimate precipitation rate and RMS precipitation-error estimates (Huffman and Bolvin 2014).

On April 15, 2015, TRMM was decommissioned after 17 years of continuously collecting data from space. Although the TMPA is still in production using the remaining satellites in the constellation, the changes and impact of the loss of

TRMM on the TMPA characteristics are expected to be small since TRMM only covers a band of 38° N-S and most of Maryland is located north of this band. In addition, the use of gauge data from GPCC will correct biases due to the loss of TRMM. The TMPA data between 1998 and 2013 are used in this study.

TMPA (Version 7) products were downloaded from the GES DISC (Liu et al. 2012). There have been few processing issues (Huffman and Bolvin 2014) before, but all the TMPA data used in this study are current.

Hurricane track data were obtained from the Best Track Data (HURDAT2), available at the NOAA National Hurricane Center (NOAA 2017b). The radius for hurricane influence is set to 500 km (Jiang and Zipser 2010; Prat and Nelson 2013) and this radius is typical for hurricanes (Unidata 2017).

When tropical cyclones make landfall, they leave moist tropical and subtropical oceans and enter drier land in mid-latitudes. As a result, water vapor as energy supply from underneath is cut off and they quickly lose strength and become remnants. When the remnants interact with frontal systems in mid-latitudes, rainfall is often enhanced when tropical warm and moist air collides with cooler and drier air from north and extra lifting is generated. Meanwhile, it is difficult to separate rainfall areas between tropical cyclones and frontal systems. In this study, we do not attempt to separate the two rain regimes when they collide together. As long as the track data are still available, rainfall within the 500-km radius is considered as hurricane rainfall (Jiang and Zipser 2010).

The terrain of Maryland is characterized with the Appalachian Mountains running through western Maryland (can obstruct passing cold fronts and cause rain shadow) and the flat area in the east adjacent to the Atlantic Ocean (exposes area to coastal weather systems).

2.4.2.2 *Preliminary Results*

From 1998 to 2013, Hurricanes Floyd, Charley, Ernesto, and Irene influenced the state of Maryland. Our preliminary results from Figure 2.11 show that Maryland experienced relatively high amounts of precipitation in the years 2003 and 2011 and relatively low amounts precipitation in the year 2001. In addition, further calculations show the average annual precipitation in Maryland is about 1,200 mm/year. Figure 2.11 shows how much precipitation that hurricanes have contributed to annual precipitation. The highest contribution of precipitation (35%) occurred in 2011, due to major hurricanes passing through Maryland such as Hurricane Irene. Additionally, in 2001 when Maryland was experiencing a drought, no hurricanes passed through the state. The average contribution of precipitation by hurricanes to annual precipitation is about 15%. Figure 2.11 shows the average of the average annual precipitation compared against each year's precipitation with and without precipitation contributed by hurricanes. Results show that when precipitation contributed by hurricanes is removed, most annual precipitations fall below the average. Some exceptions occur such as in year 2003, where Maryland was receiving relatively high monthly precipitation in general throughout the year. As a result, one could conclude that precipitation contributed by hurricanes does affect Maryland's precipitation.

2.5 SUMMARY AND FUTURE PLANS

In this chapter, we introduced NASA satellite-based data and services with emphasis on the hydrologic cycle and data products at the GES DISC. Significant progress has been made in satellite Earth's observations since the first successful launch of weather satellite, TIROS, by NASA on April 1, 1960. In particular, NASA's EOS project is a coordinated series of polar-orbiting and low inclination satellites for long-term global observations of the land surface, biosphere,

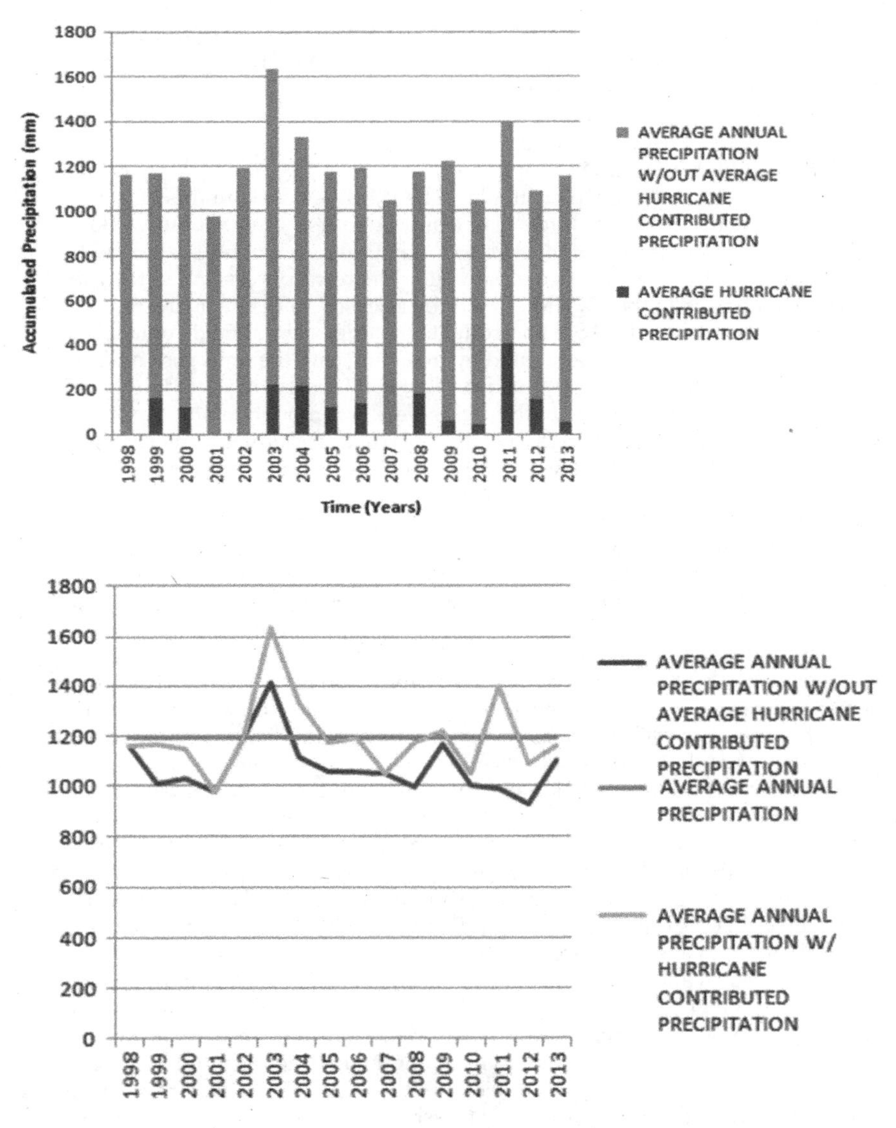

Figure 2.11 Top: Time series of annual precipitation (in mm) in Maryland with hurricane contributed precipitation highlighted. Bottom: Average annual precipitation (in red) in Maryland compared against annual precipitation with (in green) and without (in blue) hurricane contributed precipitation (units: mm).

solid Earth, atmosphere, and oceans to enable an improved understanding of the Earth as an integrated system. NASA's EOS and other past NASA satellite mission data are available and distributed by the EOSDIS, with major facilities at 12 DAACs located throughout the United States. Science disciplines at 12 DAACs include atmosphere, cryosphere, human dimensions, land, ocean, and calibrated radiance and solar radiance, etc. Datasets are emphasized and used in the examples in this article are archived at the GES DISC.

The GES DISC hosts global and regional satellite-based data products from these science disciplines: precipitation, solar irradiance, atmospheric composition and dynamics, global modeling, etc. There are over 2,700 unique data products archived at the GES DISC. In this chapter, we only present several major projects that are closely related to smart cities.

Multi-satellite and multi-sensor merged global precipitation products are available at the GES DISC. In particular, The IMERG suite contains of three output products, "Early satellites" (lag time: ~4 hours), "Late satellites" (lag time: ~12 hours), and the final "Satellite-gauge" (lag time: ~2.5 months) along with additional new input and intermediate files. The retro-processing of the IMERG product suite back to the TRMM era will be released in the near future.

Global and regional LDAS data products include optimal fields of land-surface states and fluxes. They are generated by ingesting satellite- and ground-based observational data products and using advanced land-surface modeling and data assimilation techniques. Both forcing data and model results support water resources applications, numerical weather prediction studies, numerous water and energy cycle investigations, and also serve as a foundation for interpreting satellite and ground-based observations.

The Modern-Era Retrospective analysis for Research and Applications, Version 2 (MERRA-2), has been developed at the NASA GMAO at the NASA Goddard Space Flight Center. MERRA-2 provides global data beginning in 1980 and runs a few weeks behind real-time. Alongside the meteorological data assimilation using a modern satellite database, MERRA-2 includes an interactive analysis of aerosols that feed back into the circulation, uses NASA's observations of stratospheric ozone and temperature (when available), and takes steps toward representing cryogenic processes. The MERRA project focuses on historical analyses of the hydrological cycle on a broad range of weather and climate time scales and places the NASA EOS suite of observations in a climate context.

Two popular point-and-click tools are presented. First, NASA's Worldview provides the capability to interactively browse global, full-resolution satellite imagery and then download the underlying data (NASA 2017g). Worldview contains 400+ NASA satellite-based products and most of them are updated within three hours of observation, essentially showing the entire Earth as it looks "right now" (NASA 2017g). On the other hand, NASA's GES DISC Giovanni is designed for in-depth data analysis and visualization. There are 8 disciplines and 74 measurements available in Giovanni and they are listed in Tables 2.3 and 2.4, respectively. Over 1,900 variables are currently available in Giovanni as of this writing and more are being added. Data variables in Giovanni are multi-disciplinary and users can access them without downloading data and software (Liu and Acker 2017). Over the years, a wide range of activities of using Giovanni has been reported by users, ranging from classroom activities to scientific investigation. Over 1,700 peer-reviewed papers across various Earth science disciplines and other areas were published with help from Giovanni.

NASA satellite-based data products at the GES DISC are also accessible (NASA 2017i) via other Web services and protocols including https (the data archive), OPeNDAP, WMS, GDS, etc.

Two examples were presented regarding the use of satellite-based data products in understanding environment changes and conditions in megacities. In Example 1, the TMPA precipitation dataset was presented with the accumulated rainfall map for Typhoon Nida that made landfall in the PRD region. Time series plots of several atmospheric composition datasets from MERRA-2 were plotted and analyzed, showing significant changes in the atmospheric environment in PRD, which could be associated with the unprecedented economic growth and urbanization in the region, especially since 2000. It is necessary to mention that these preliminary results need to be verified independently with ground or other measurements to address biases and other issues.

In Example 2, the effect of hurricanes on annual precipitation in Maryland was investigated. From 1998 to 2013, Hurricane Floyd, Charley, Ernesto, and Irene influenced the state of Maryland. Our preliminary results from Figure 2.11 show that Maryland experienced relatively high amounts of precipitation in years 2003 and 2011 and relatively low amounts precipitation in the year 2001. In addition, further calculations show the average annual precipitation in Maryland is about 1,200 mm/year. Figure 2.11 shows how much precipitation hurricanes have contributed to annual precipitation. The highest contribution of precipitation (35%) occurred in 2011, possibly due to major hurricanes passing through Maryland such as Hurricane Irene. Additionally, in 2001 when Maryland was experiencing a drought, no hurricanes passed through the state. The average contribution of precipitation by hurricanes to annual precipitation is about 15%. Figure 2.11 shows the average of the average annual precipitation compared against each year's average precipitation with and without precipitation contributed by hurricanes. Results show that when precipitation contributed by hurricanes is removed, most annual precipitations fall below the average for each. Some exceptions occur such as in year 2003, where Maryland was receiving relatively high monthly precipitation in general. As a result, one could conclude that precipitation contributed by hurricanes does affect Maryland's precipitation.

As of this writing, EOSDIS hosts ~22 PB of EO data at 12 DAACs and it is expected to grow to more than 37 (246 PB) PB by 2020 (2025) (NASA 2017a). Such large EO data archive is an important asset for environmental research and applications around the world. Over the years, NASA EOSDIS has developed data services and tools to facilitate data discovery and access in twelve discipline-oriented DAACs. For complex issues of smart cities, it often requires a multi-disciplinary approach which needs an information system that can integrate all these data products archived at twelve DAACs as well as other related data products from users, providing a one-stop shop for data and services by removing obstacles such as data discovery, access, interoperability, etc., and better address environmental issues encountered in city planning, operations, research, etc. (Güneralp and Seto 2008). The NASA GES DISC Giovanni is an example that makes multi-discipline data variables available in one place and efforts are being carried out to include additional datasets from other DAACs and make them available in Giovanni. The ongoing NASA EOSDIS Cloud Evolution Project (NASA 2017j) will have the potential for developing an information system that supports multi-disciplinary data products and services. Moving from discipline-oriented to multi-discipline-oriented data services is not a simple task, which will involve in a team of data scientists from NASA and other organizations as well as from end users of smart cities due to many obstacles to be overcome such as data formats, data volume, data structures, terminology in different disciplines, etc. Nonetheless, still a lot of work needs to be done to develop better information systems and services for efficiently solving problems in smart cities.

ACKNOWLEDGMENTS

We recognize the team effort of all past and current members at the GES DISC for their contributions to the development of data services and tools such as Giovanni. We extend our thanks to data set algorithm developers and many users for their feedback and suggestions. The GES DISC is funded by NASA's Science Mission Directorate.

REFERENCES

Acker, J. G. and G. Leptoukh, 2007: Online Analysis Enhances Use of NASA Earth Science Data. *Eos. Trans. Amer. Geophys. Union*, 88 (2), 14–17.

Adler, R.F., G.J. Huffman, A. Chang, R. Ferraro, P. Xie, J. Janowiak, B. Rudolf, U. Schneider, S. Curtis, D. Bolvin, A. Gruber, J. Susskind, P. Arkin, and E. Nelkin 2003: The Version 2 Global Precipitation Climatology Project (GPCP) Monthly Precipitation Analysis (1979-Present). *J. Hydrometeor.*, 4, 1147–1167.

Aonashi, K., J. Awaka, M. Hirose, T. Kozu, T. Kubota, G. Liu, S. Shige, S. Kida, S. Seto, N. Takahashi, and Y. N. Takayabu, 2009: GSMaP Passive, Microwave Precipitation Retrieval Algorithm: Algorithm Description and Validation. *J. Meteor. Soc. Japan*, 87A, 119–136.

Behrangi, A., K.-L. Hsu, B. Imam, S. Sorooshian, G. J. Huffman, and R. J. Kuligowski, 2009: PERSIANN-MSA: A Precipitation Estimation Method from Satellite-Based Multispectral Analysis. *J. Hydrometeor*, 10, 1414–1429. doi: 10.1175/2009JHM1139.1.

Bitew, M., M., M. Gebremichael, L. T. Ghebremichael, and Y. A. Bayissa, 2012: Evaluation of High-Resolution Satellite Rainfall Products Through Streamflow Simulation in a Hydrological Modeling of a Small Mountainous Watershed in Ethiopia. *J. Hydrometeor*, 13, 338–350. doi: 10.1175/2011JHM1292.1.

Bosilovich, M. G., F. R. Robertson, and J. Chen, 2011: Global energy and water budgets in MERRA. *132 J. Climate*, 24, 282–300.

Bosilovich, M., S. Akella, L. Coy, R. Cullather, C. Draper, R. Gelaro, R. Kovach, Q. Liu, A. Molod, P. Norris, K. Wargan, W. Chao, R. Reichle, L. Takacs, Y. Vikhliaev, S. Bloom, A. Collow, S. Firth, G. Labow, G. Partyka, S. Pawson, O. Reale, S. Schubert, and M. Suarez, 2015: Technical Report Series on Global Modeling and Data Assimilation, Vol. 43, R. Koster, Editor. Available online: http://gmao.gsfc.nasa.gov/pubs/tm/docs/Bosilovich803.pdf.

Boucher, A. and K. C. Seto, 2009: Methods and Challenges for Using High-Temporal Resolution Data to Monitor Urban Growth. *Global Mapping of Human Settlements: Experiences, Data Sets, and Prospects.* P. Gamba and M. Herold, Editors. Boca Raton, FL: CRC Press, 2009. 339+.

Census Bureau, 2016: Four Texas Metro Areas Collectively Add More Than 400,000 People in the Last Year, Census Bureau Reports. Available online: www.census.gov/newsroom/press-releases/2016/cb16-43.html, last accessed September 6, 2017.

Engel T., Fink A.H., Knippertz P., Pante G., and Bliefernicht J., 2017. Extreme Precipitation in the West African Cities of Dakar and Ouagadougou-Atmospheric Dynamics and Implications for Flood Risk Assessments. *J. Hydrometeorol.*, 18, 2937–2957.

Gallaher, D. and G. Grant, 2012: Data Rods: High Speed, Time-Series Analysis of Massive Cryospheric Data Sets Using Pure Object Databases. Geoscience and Remote Sensing Symposium (IGRASS), 22–27 July 2012. Available online: http://ieeexplore.ieee.org/xpl/articleDetails.jsp?reload=true&arnumber=6352413&contentType=Conference+Publications.

GDS, 2017, The GrADS Data Server (GDS). Available online: http://cola.gmu.edu/grads/gds.php, last accessed September 6, 2017.

Gianotti, R. L., D. Zhang, and E. A. B. Eltahir, 2012: Assessment of the Regional Climate Model Version 3 Over the Maritime Continent Using Different Cumulus Parameterization and Land Surface Schemes. *J. Climate*, 25, 638–656. doi: 10.1175/JCLI-D-11-00025.1.

Gourley, J. J., Y. Hong, Z. L. Flamig, J. Wang, H. Vergara, and E. N. Anagnostou, 2011: Hydrologic Evaluation of Rainfall Estimates from Radar, Satellite, Gauge, and Combinations on Ft. Cobb Basin, Oklahoma. *J. Hydrometeor*, 12, 973–988. doi: 10.1175/2011JHM1287.1.

Güneralp, B. and K. C. Seto, 2008. Environmental Impacts of Urban Growth from an Integrated Dynamic Perspective: A Case Study of Shenzhen, South China. *Global Environ. Change*, 18(4), 720–735. doi: 10.1016/j.gloenvcha.2008.07.004.

Guo, J. M. Deng, S. S. Lee, F. Wang, Z. Li, P. Zhai, H. Liu, W. Lv, W. Yao, and X. Li, 2016: Delaying Precipitation and Lightning by Air Pollution Over Pearl River Delta. Part I: Observational Analyses. *J. Geophys. Res. Atmos.*, 121, 6472–6488. doi: 10.1002/2015JD023257.

Hong, Y., D. Gochis, J. Cheng, K.-L. Hsu, and S. Sorooshian, 2007: Evaluation of PERSIANN-CCS Rainfall Measurement Using the NAME Event Rain Gauge Network. *J. Hydrometeor*, 8, 469–482. doi: 10.1175/JHM574.1.

Huffman, G.J, R.F. Adler, D.T. Bolvin, and G. Gu, 2009: Improving the Global Precipitation Record: GPCP Version 2.1. *Geophys. Res. Lett.*, 36, L17808. doi:10.1029/2009GL040000.

Huffman, G.J., R.F. Adler, D.T. Bolvin, G. Gu, E.J. Nelkin, K.P. Bowman, Y. Hong, E.F. Stocker, and D.B. Wolff, 2007: The TRMM Multi-Satellite Precipitation Analysis: Quasi-Global, Multi-Year, Combined-Sensor Precipitation Estimates at Fine Scale. *J. Hydrometeor.*, 8(1), 38–55.

Huffman, G.J., R.F. Adler, D.T. Bolvin, and E.J. Nelkin, 2010: The TRMM Multi – Satellite Precipitation Analysis (*TAMPA*). Chapter 1 in *Satellite Rainfall Applications for Surface Hydrology*. F. Hossain and M. Gebremichael, Editors, Springer Dordrecht Heidelberg London New York, ISBN: 978-90-481-2914-0, 3-22.

Huffman, G.J. and D.T. Bolvin, 2012: Real-Time TRMM Multi-Satellite Precipitation Analysis Data Set Documentation. Available online: ftp://trmmopen.gsfc.nasa.gov/pub/merged/V7Documents/3B4XRT_doc_V7.pdf, last accessed June 8, 2014.

Huffman, G.J. and D.T. Bolvin, 2013: TRMM and Other Data Precipitation Data Set Documentation. Available online: ftp://meso-a.gsfc.nasa.gov/pub/trmmdocs/3B42_3B43_doc.pdf, last accessed September 6, 2017.

Huffman, G.J. and D.T. Bolvin, 2014: TRMM and Other Data Precipitation Data Set Documentation. Available online: ftp://meso-a.gsfc.nasa.gov/pub/trmmdocs/3B42_3B43_doc.pdf, last accessed September 6, 2017.

Huffman, G.J., D. Bolvin, D. Braithwaite, K. Hsu, R. Joyce, C. Kidd, E. Nelkin, S. Sorooshian, J. Tan, and P. Xie, 2017: IMERG Algorithm Theoretical Basis Document (ATBD). Available online: https://pmm.nasa.gov/sites/default/files/document_files/IMERG_ATBD_V4.6.pdf, last accessed September 6, 2017.

Jiang, H. and E. J. Zipser, 2010: Contribution of Tropical Cyclones to the Global Precipitation from Eight Seasons of TRMM Data: Regional, Seasonal, and Interannual Variations. *J. Climate*, 23, 1526–1543. doi: 10.1175/2009JCLI3303.1.

Jin, M., W. Kessomkiat, and G. Pereira, 2011: Satellite-Observed Urbanization Characters in Shanghai, China: Aerosols, Urban Heat Island Effect, and Land-Atmosphere Interactions. *Remote Sensing*, 3, 83–99. doi: 10.3390/rs3010083.

Jin, M., J. M. Shepherd, and M. D. King, 2005: Urban Aerosols and Their Interaction with Clouds and Rainfall: A Case Study for New York and Houston. *J. Geophysical Research*, 110, D10S20. doi: 10.1029/2004JD005081.

Jin, M., J. M. Shepherd, and W. Zheng, 2010: Urban Surface Temperature Reduction via the Urban Aerosol Direct Effect – A Remote Sensing and WRF Model Sensitivity Study PDF file of the paper Advances in Meteorology, Volume 2010, Article ID 681587, 14 pages. doi: 10.1155/2010/681587.

Joyce, R. J., J. E. Janowiak, P. A. Arkin, and P. Xie, 2004: CMORPH: A Method That Produces Global Precipitation Estimates from Passive Microwave and Infrared Data at High Spatial and Temporal Resolution. *J. Hydromet.*, 5, 487–503.

Kaufmann, R. K., K. C. Seto, A. Schneider, L. Zhou, Z. Liu, and W. Wang, 2007: Climate Response to Rapid Urban Growth: Evidence of a Human-Induced Precipitation Deficit. *J. Climate*, 20(10), 2299–2306. doi: 10.1175/JCLI4109.1.

Kearns, E. 2017: Improving Access to Open Data through NOAA's Big Data Project. Available online: https://bigdatawg.nist.gov/Day2_08_NIST_Big_Data-Kearns.pdf, last accessed September 6, 2017.

Lee, T., S. Miller, F. Turk, C. Schueler, R. Julian, S. Deyo, P. Dills, and S. Wang, 2006: The NPOESS VIIRS Day/Night Visible Sensor. *Bull. Amer. Meteor. Soc.*, 87, 191–199. doi: 10.1175/BAMS-87-2-191.

Liu, J. and Z. Liu, 2015: The Connection Between Hurricanes and Precipitation in Maryland. *Global Precipitation Measurement, Validation, and Applications III, AGU Fall Meeting*, San Francisco, December 14–18, 2015.

Liu, Z. and J. Acker, 2017: Giovanni: The Bridge Between Data and Science. *Eos*, 98. doi: 10.1029/2017EO079299. Published on August 24, 2017.

Liu, Z., D. Ostrenga, W. Teng, and S. Kempler, 2012: Tropical Rainfall Measuring Mission (TRMM) Precipitation Data and Services for Research and Applications. *Bull. Am. Meteorol. Soc.* 93, 1317–1325. doi:10.1175/BAMS-D-11-00152.1.

Liu, Z., D. Ostrenga, B. Vollmer, et al. 2017: Global Precipitation Measurement Mission Products and Services at the NASA GES DISC. *Bull. Am. Meteorol. Soc.*, 98(3), 437–444 [10.1175/bams-d-16-0023.1].

Mahrooghy, M., V. G. Anantharaj, N. H. Younan, J. Aanstoos, and K.-L. Hsu, 2012: On an Enhanced PERSIANN-CCS Algorithm for Precipitation Estimation. *J. Atmos. Oceanic Technol.*, 29, 922–932. doi: 10.1175/JTECH-D-11-00146.1.

NASA, 2017a, NASA's Earth Observing System Project Science Office. Available online: https://eospso.gsfc.nasa.gov/, last accessed September 6, 2017.

NASA, 2017b, NASA Earth Science Data and Information System Project. Available online: https://earthdata.nasa.gov/about/esdis-project, last accessed September 6, 2017.

NASA, 2017c: Land Data Assimilation Systems. Available online: https://ldas.gsfc.nasa.gov/gldas/, last accessed September 6, 2017.

NASA, 2017d: Modern-Era Retrospective analysis for Research and Applications, Version 2. Available online: https://gmao.gsfc.nasa.gov/reanalysis/MERRA-2/, last accessed September 6, 2017.

NASA, 2017e: Earthdata Search. Available online: https://search.earthdata.nasa.gov/search, last accessed September 6, 2017.

NASA, 2017f: EOSDIS Data Tools. Available online: https://earthdata.nasa.gov/earth-observation-data/tools, last accessed September 6, 2017.

NASA, 2017g: NASA Worldview. Available online: https://worldview.earthdata.nasa.gov/, last accessed September 6, 2017.

NASA, 2017h: NASA GES DISC Giovanni. Available online: https://giovanni.gsfc.nasa.gov/, last accessed September 6, 2017.

NASA, 2017i: Web Services at GES DIC. Available online: https://disc.gsfc.nasa.gov/information/tools?title=OPeNDAP%20and%20GDS, last accessed September 6, 2017.

NASA, 2017j: EOSDIS Cloud Evolution. Available online: https://earthdata.nasa.gov/about/eosdis-cloud-evolution, last accessed September 6, 2017.

NOAA, 2017a: Historical Hurricane Tracks. Available online: https://csc.noaa.gov/hurricanes/, last accessed September 6, 2017.

NOAA, 2017b: Hurricane Best Track Data. Available online: www.nhc.noaa.gov/data/, last accessed September 6, 2017.

OPeNDAP, 2017: OPeNDAP – Advanced Software for Remote Data Retrieval. Available online: https://www.opendap.org/, last accessed September 6, 2017.

Prat, O. P. and B. R. Nelson, 2013: Precipitation Contribution of Tropical Cyclones in the Southeastern United States from 1998 to 2009 Using TRMM Satellite Data. *J. Climate*, 26, 1047–1062. doi: 10.1175/JCLI-D-11-00736.1.

Reichle, R. H., and Q. Liu, 2014: Observation-Corrected Precipitation Estimates in GEOS-5. NASA/TM–2014-104606, Vol. 35.

Reichle, R. H., R. D. Koster, G. J. M. De Lannoy, B. A. Forman, Q. Liu, S. Mahanama, and A. Toure, 2011: Assessment and Enhancement of MERRA Land Surface Hydrology Estimates. *J. Climate*, 24, 6322–6338, doi: 10.1175/JCLI-D-10-05033.1.

Rienecker, M.M., M.J. Suarez, R. Gelaro, R. Todling, J. Bacmeister, E. Liu, M.G. Bosilovich, S.D. Schubert, L. Takacs, G.-K. Kim, S. Bloom, J. Chen, D. Collins, A. Conaty, A. da Silva, et al. 2011: MERRA: NASA's Modern-Era Retrospective Analysis for Research and Applications. *J. Climate*, 24, 3624–3648. doi: 10.1175/JCLI-D-11-00015.1.

Rui, H., R. Strub, W.L. Teng, B. Vollmer, D.M. Mocko, D.R. Maidment, and T.L. Whiteaker, 2013: Enhancing Access to and Use of NASA Earth Sciences Data Via CUAHSI-HIS (Hydrologic Information System) and Other Hydrologic Community Tools. *AGU Fall Meeting*, San Francisco, CA, December 9–13, 2013.

Rui, H., B. Teng, R. Strub, and B. Vollmer, 2012: Data Reorganization for Optimal Time Series Data Access, Analysis, and Visualization. AGU Fall Meeting, San Francisco, CA, December 3–7, 2012.

Seto, K. C., 2007. Urbanization in China: The Pearl River Delta Example. *Our Changing Planet: The View From Space*. M. D. King, C. L. Parkinson, K. C. Partington and R. G. Williams, Editors. Cambridge: Cambridge University Press.

Seto, K. C., 2011: Monitoring Urban Growth and Its Environmental Impacts Using Remote Sensing: Examples from China and India. *Global Urbanization*. E. Birch and S. Wachter, Editors. Philadelphia: University of Pennsylvania Press, 2011. 151–166.

Seto, K. C. and J. M. Shepherd, 2009: Global Urban Land-Use Trends and Climate Impacts. *Curr. Op. Environ. Sustain.*, 1(1), 89–95. doi: 10.1016/j.cosust.2009.07.012.

Shepherd, M. J. and M. Jin, 2004: Linkages Between the Urban Environment and Earth's Climate System. *EOS*, 85, 227–228.

Su, F., H. Gao, G. J. Huffman, and D. P. Lettenmaier, 2011: Potential Utility of the Real-Time TMPA-RT Precipitation Estimates in Streamflow Prediction. *J. Hydrometeor*, 12, 444–455. doi:10.1175/2010JHM1353.122.

Suarez, M. and J. Bacmeister, 2015: Development of the GEOS-5 Atmospheric General Circulation Model: Evolution from MERRA to MERRA2. *Geosci. Model Dev.*, 8, 1339–1356. doi: 10.5194/gmd-8-1339-2015.

Takacs, L. L., M. Suarez, and R. Todling, 2015. Maintaining Atmospheric Mass and Water Balance Within Reanalysis. NASA/TM–2014-104606, Vol. 37.

Tan, M.L. and Duan, Z. 2017: Assessment of GPM and TRMM Precipitation Products Over Singapore. *Remote Sens.*, 9, 720.

Tekeli, A.E., 2017: Exploring Jeddah Floods by Tropical Rainfall Measuring Mission Analysis. *Water*, 9, 612.

Teng, W., H. Rui, R. Strub, and B. Vollmer, 2016: Optimal Reorganization of NASA Earth Science Data for Enhanced Accessibility and Usability for the Hydrology Community. *J. Am. Water Resource Assoc. (JAWRA)*, Vol. 52, Issue 4, 825–835. doi: 10.1111/1752-1688.12405.

Unidata, 2017: Hurricane Structure - Excerpt from: Community Hurricane Preparedness. Available online: https://www.unidata.ucar.edu/data/NGCS/lobjects/chp/structure/, last accessed September 6, 2017.

United Nations, 2014: World Urbanization Prospects. Available online: https://esa.un.org/unpd/wup/Publications/Files/WUP2014-Highlights.pdf.

Wikipedia, 2017a, The Pearl River Delta. Available online: https://en.wikipedia.org/wiki/Pearl_River_Delta, last accessed September 6, 2017.

Wikipedia, 2017b: Tropical Storm Nida (2016). Available online: https://en.wikipedia.org/wiki/Tropical_Storm_Nida_(2016), last accessed, September 6, 2017.

Wikipedia, 2017c: Pearl River Delta Economic Zone. Available online: https://en.wikipedia.org/wiki/Pearl_River_Delta_Economic_Zone, last accessed September 6, 2017.

WMS, 2017: Web Map Service. Available online: https://en.wikipedia.org/wiki/Web_Map_Service, last accessed September 6, 2017.

Wu H., R. F. Adler, Y. Hong, Y. Tian, and F. Policelli, 2012: Evaluation of Global Flood Detection Using Satellite-Based Rainfall and a Hydrologic Model. *J. Hydrometeor*, 13, 1268–1284.

Zhang, H., M. Jin, and M. Leach, 2017: A Study of the Oklahoma City Urban Heat Island Effect Using a WRF/Single-Layer Urban Canopy Model, a Joint Urban 2003 Field Campaign, and MODIS Satellite Observations. *Climate*, 5(3), 72; doi: 10.3390/cli5030072

3 Advancing Smart and Resilient Cities with Big Spatial Disaster Data

Challenges, Progress, and Opportunities

Xuan Hu and Jie Gong

CONTENTS

3.1 INTRODUCTION

Nearly 500 natural disasters occur each year in the world. While the causes of them vary from event to event, what they share in common is that each time they strike, these unexpected events cause loss of lives, profound economic loss, social disruption, and so. Coastal communities possess one of the most dynamic interfaces between human civilization and the natural environment. In the worldwide, over 38% of the human population lives in the coastal zone, with over half of the human population live in coastal counties in the United States. The coastal communities in the United States have been recently severely impacted by several major hurricane events including Hurricane Katrina (2005), Hurricane Sandy (2012), Hurricane Arthur (2014), and Hurricane Mathews (2016). For example, Hurricane Sandy (2012) wreaked havoc and destruction across the Atlantic coastal area spanning 24 states, flooding and destroying residential houses, knocking down power lines systems, paralyzing transportation system – not to mention the shattered lives of hundreds of thousands of people.

With the climate system becoming increasingly aggressive, future storm events are likely to raise the cost in coastal areas in a variety of ways. First, the impacts of climate change are likely to worsen the problems that coastal areas already face. Sea level rise, increase in precipitation, and changes in frequency and intensity of storm events are increasingly exposing the vulnerability of coastal areas. Second, future coastal development may reduce the resilience of natural systems to respond to storm events. Increase in population, construction of physical infrastructures as well as growing human activities are prone to disrupt natural coastal and marine ecosystems. Eventually, the fundamental weakness of humankind comes to the fore every time a storm comes, and the formidable force of nature topples the strongest of a person.

To cope with the continuing storm events and the inherent vulnerabilities in coastal communities, enhancing community resilience is essential. Building resilience requires improvement in "adaptability." Improving the "adaptability" of community resilience needs real-time situation awareness to deal with dynamic threats. It is critical for experts, decision-makers, and emergency personnel to have access to real-time information in order to assess the situation and respond appropriately (NSF 2015). The need for information is intensified in these small probability events because predicting the impact of these events is under normal circumstance impossible using either human experience or historical data. Han et al. (1998) highlight the importance of real-time information. They stated that to cope with disasters, a system needs to modify its strategy immediately according to real-time disaster situations. Nevertheless, disaster scenarios are under normal circumstance impossible to be fully enumerated by domain experts without rapidly updated information.

While there is a global consensus on the need for timely information, the merit of spatial data is not fully recognized until recently as more spatial data are being collected and analyzed during extreme events. Though the phrase "80% of data is geographic" is arguable (Garson and Biggs 1992), most disaster-related data have some kind of geospatial features. Due to the lack of efficient methods to extract information from these data sets, traditional ways of using these data sets often give up the spatial components and attempt to tailor the data to transactional database structure despite the fact that these spatial components may contain essential information for decision-making.

There are fundamental challenges in managing, analyzing, and interpreting the growing size and complexity of spatial disaster datasets. The vast size and complex processing requirements of these new data sets make it challenging to utilize effectively in real-life scenarios. For example, during large-scale coastal storm events, crucial information is often hidden in these data sets and is in no way integrated into ongoing decision processes. To fully exploit their potential, we need to revisit existing spatial data analytics and develop new capabilities to rapidly synthesize information from data at rest (i.e., data already stored in the system) and data-in-transit from sensors and make information available on time and at the relevant level of decision-making. This study uses immense quantities of spatial disaster data collected during Hurricane Sandy as the empirical datasets to characterize the anatomy of big spatial disaster data and to analyze the processing patterns in using these data sets in disaster management applications. This chapter addresses the following critical research questions: (1) What is the basic anatomy of spatial disaster data; and (2) What are the core operation categories and processing patterns with spatial disaster data? The answers to these questions advance our understanding on the basic principles and processes involved in big spatial disaster data-driven disaster response and recovery as well as the critical research needs for improving data support for disaster recovery decision-making.

3.2 THE ROLE OF SPATIAL DATA IN COASTAL RESILIENCE APPLICATIONS

3.2.1 Disaster Management Cycle

In general, disaster management, structured as disaster management cycle in Figure 3.1, consists of four phases including mitigation, preparation, response, and recovery. Although, the cycle can be considered as a continuum (Joyce et al. 2009). Usually, the mitigation phase is considered as the first phase. Mitigation incorporates all possible activities to enhance the inherent "capability" of a community through reducing the chance of a hazard from happening, preventing a hazard from forming a disaster, and minimizing the damaging effects of the inevitable disaster. This is achieved through the process of risk identification, assets structure modification, etc. Preparedness is another phase that emphasizes on development of community "capability" to deal with the potential hazard. This phase includes readiness planning and activities to handle a disaster in the realization of residual risks. Objectives in the preparedness phase include identification and development of necessary systems, skills, and resource before hazards occur. The next phase is response phase, which is closely relevant to the "adaptability" behavior of a community. This phase focuses primarily on enforcing operations to protect life and property during disasters. Typical activities during the response phase include evacuation, search and rescue, establishing immediate emergency shelters, etc. Finally, recovery is the phase that deals with the aftermath of a disaster. Recovery phase includes both short-term tasks such as restoration of lifeline essentials and the longer-term tasks such as rebuilding of communities. Moreover, the recovery phase can be considered as the

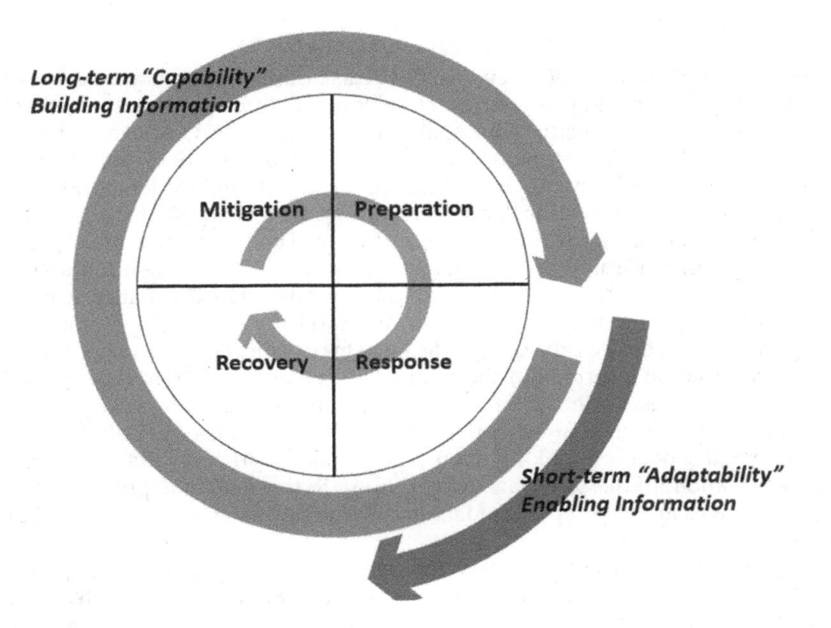

Figure 3.1 Disaster management cycle.

Table 3.1: Key Categories of Information Needed in Disaster Management

Category	Applicable Phases	Key Features	Objective
Long-term "capability" building information	Mitigation, Preparation, Recovery & response	Resolution sensitive and spatially relevant	Provide baseline information for risk identification, damage impact prediction, community system monitoring, community recovery planning, and disaster response
Short-term "adaptability" enabling information	Response	Time-sensitive, spectrally-relevance, acceptable spatial relevance	Provide immediate damage assessment information for disaster operations

summary of a previous disaster that will have far-reaching implication in building the long-term "capability" of the next coming hazard (Becker et al. 2008).

What these four disaster management phases share in common is that they all need some kind of information and knowledge to reinforce corresponding actives and operations. In today's increasingly networked, digitized, sensor-laden, information-driven world, big data provide new insights into disaster management. The objective of using big data in a disaster is to provide the appropriate information in a temporally, and spatially relevant context (Joyce et al. 2009). In the light of this, big data need to be adapted to meet the requirements of all four phases. In general, the need for information in disaster management cycle can be divided into two aspects (Table 3.1): long-term "capability" building information and short-term "adaptability" enabling information.

3.2.2 Data Acquisition

To cope with the information need for the disaster management cycle, different data acquisition and processing technologies are often deployed to gather information. Baseline ancillary geospatial data are historical geospatial data that previously collected and stored in transactional databases. These data provide essential baseline information including but are not limited to, demography information, terrain elevation, land use, building footprints (Figure 3.2), critical infrastructure information (Figure 3.3), and so on. The value of ancillary geospatial data in resilience analysis is widely recognized such as in flood inundation modeling (Sanders 2007) and behavior models for disasters evacuation (Chen et al. 2011). However, these data are static and carefully prepared for general purposes: they sacrifice processing efficiency for details, and they do not convey real-time situation information about the impacted areas. Therefore, under normal circumstances, these ancillary spatial data are potent in predictive analysis, but not equally effective in validation or interactive analyses. Validation models and interactive decision-making often urges the acquisition of data with high frequency and rapidness (Miyazaki et al. 2015). To this end, many remote sensing techniques are now capable of facilitating the capturing of real-time spatial data.

Imagery data is one of the most common types of data collected during natural disasters. Real-time or near real-time observations from satellite, aerial, and ground platforms serve as essential means for imagery data collection for enhancing resilience (Gillespie et al. 2007), vulnerability analysis (Walker 1996),

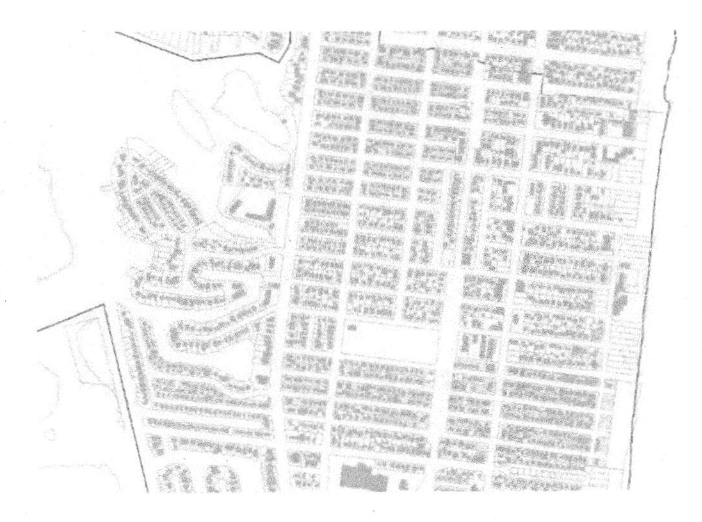

Figure 3.2 Ancillary geospatial data showing building outline polygons.

Figure 3.3 Ancillary geospatial data showing main and distributed pipeline network.

and decision support (Voigt et al. 2007). Satellite imagery data (Figure 3.4) is most efficient in terms of capturing the terrain condition of a spacious area (Miyazaki et al. 2015). Since the repeat interval of most satellites is often daily to monthly (Gillespie et al. 2007), it is most suitable for monitoring and modeling relationship between human activity and long-term environmental or climate impact (Miyazaki et al. 2015). Compared to the lengthy revisit time of satellite imagery methods, aircrafts or drones are often ready to be deployed to capture either vertical or oblique imagery (Figure 3.5) in a timely manner. Such rapidness contributes to the prompt situation awareness in early impact analysis (Hirokawa et al. 2007). Meanwhile, oblique imagery often has a higher resolution

Figure 3.4 Satellite Imagery in Ortlety Beach after Hurricane Sandy. (Adopted from Google Earth.)

a. Oblique Imagery in Ortley Beach Before Hurricane Sandy

b. Oblique Imagery in Ortley Beach After Hurricane Sandy

Figure 3.5 Oblique Imagery in Ortley (a) before and (b) after Hurricane Sandy (Data from USGS).

than the satellite imagery data, making it more accurate for supporting post-disaster assessment (Ezequiel et al. 2014).

Besides 2D image processing applications such as segmentation (Shi and Malik 2000), another process called photogrammetry (Figure 3.6) attempt to extract the third dimension or depth information from the 2D images (Zhu and Brilakis 2007). It has been adopted to recovery the 3D spatial information of buildings (Suveg and Vosselman 2004), infrastructures (Brilakis et al. 2011), underwater structures (Beall et al. 2010), and indoor scenes (Izadi et al. 2011). Furthermore, a method derived from photogrammetry method is videogrammetry in which cameras continuously record video frames in a given time span and the sequential characteristic such a recording mechanism allows feature points in consecutive frames to match automatically. Notwithstanding the advantages of using imagery-based data, they have their limitations. In particular, the quality of imagery-based information analysis is sensitive to environmental conditions such as light, weather or relative positions as well as the functionality of processing software programs (Dai and Lu 2010). Suitability of such techniques in terms of spatial accuracy is still a major concern (Bhatla et al. 2012).

Light Detection and Ranging (LiDAR) is another emerging technology that facilitates the collection of spatial data in a disaster environment. This technology is built on the principle of the time of flight: a laser scanner automatically records the scanning angels and the travel time (from the emission of a pulse to its return) of each probed object. Hundreds of thousands of point measurements in the form of x, y, z coordinates and intensity values can be saved each second, and accumulatively forming 3D shapes as point clouds. LiDAR data have multiple benefits. First, LiDAR uses active sensing mechanisms, eliminating the need for ambient light to operate. As a result, it can be functional at night time or in other not so ideal light conditions. Second, it provides data in better spatial accuracy compared to imagery approaches (Csanyi and Toth 2007). Third, the LiDAR instruments can be mounted onto different platforms such as mobile and aerial platforms, and their data coverage rate can reach hundreds of miles per hour. On account of these three advantages, LiDAR technologies accommodate the needs of mapping both large-scale objects such as

Figure 3.6 3D Reconstruction using 2D imagery data (Adopted from Zhou et al. (2015)). (a) High-resolution 2D imagery data by ground survey team. (b) 3D model (SURE) reconstructed based on the high-resolution 2D imagery. (c) 3D model (123D CATCH) reconstructed based on the high-resolution 2D imagery.

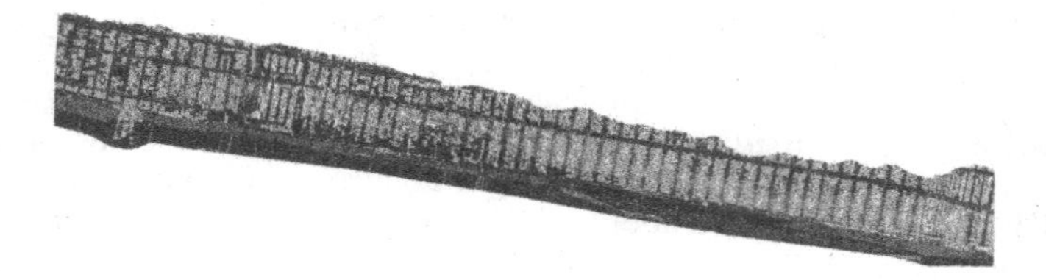

Figure 3.7 Airborne LiDAR data (color by classification, USGS).

terrain (Lefsky et al. 2002) (Figure 3.7) and small-scale objects such as buildings and vegetation (Gong et al. 2012). This technology has been widely deployed in earthquakes (Olsen et al. 2012), flooding (Poulter and Halpin 2008) forest fires (Morsdorf et al. 2004), and hurricanes (Gong et al. 2012). Apart from these benefits, one significant concern with LiDAR data is their complex and intensive computation needs in data indexing and storage (Schön et al. 2009), visualizations, and feature extraction (Rottensteiner and Briese 2002). The complexity of these algorithms often results in lengthy computation time, which will eventually hamper the effective utilization of LiDAR data.

More recently, mobile devices, as well as advances in social media, have provided new paths for citizens or individuals to generate non-expert spatial data. Average citizens, even without domain knowledge, are able to record geotagged disaster situations (Goodchild and Glennon 2010) and contribute to the collection of a new type of spatial disaster data called Volunteered Geographic Information (VGI). This has increasingly become a primary communication channel between citizens and public authorities (Miyazaki et al. 2015). It empowers authorities or experts to hear from "people's voice" though "crowd voting." "People's voice" provides a unique gateway for decision-makers and specialists to aware the disaster situation through feeling sentiments of victims. For example, by analyzing the tweets, researchers are able to promptly identify the disaster-prone area (Earle et al. 2012) and conduct early impact analysis (Sakaki et al. 2010). Despite the merit of using crowdsourcing data, the challenge remains significant. Because, these spatial data are from non-authority sources, the quality of such data remains questionable (Allahbakhsh et al. 2013). For example, researchers have studied the impacts of fake and incorrect information during Hurricane Sandy (Gupta et al. 2013). They concluded that the artificial error could pose significant threats to the credibility of the "crowd voting."

3.2.3 Challenges and Opportunities

Geospatial data have frequently been used in vulnerability assessment, which is closely related to identifying gaps in a community's capabilities to cope with extreme events. Although the use of emerging large geospatial data sets in these types of analysis is difficult, the analysis can be eventually accomplished given sufficient time. What most challenging is to use these data sets to obtain better situation awareness in time-sensitive applications. Current geospatial data analysis frameworks are inadequate in handling these large datasets, especially during large-scale extreme events.

Figure 3.8 depicts a simplified map of current data analytics for disaster response practice that essentially involves three key processes including data acquisition, data processing, and decision-making. In the first step, data are collected from multiple resources such as mobile platforms, airborne systems, and social media outlets. The next step is data processing, which analyzes the data according to the particular goals such as detecting the morphology changes of

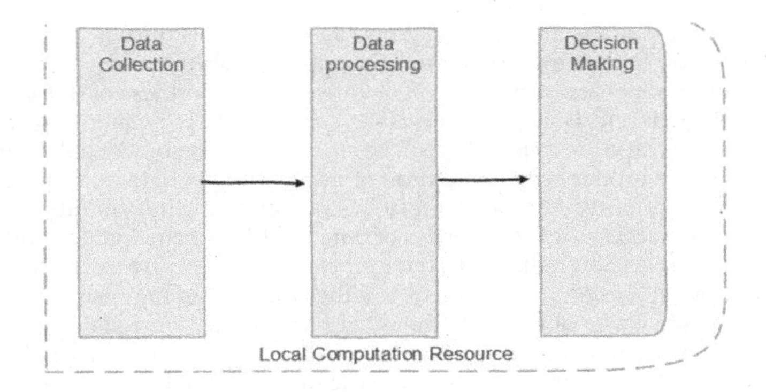

Figure 3.8 A simplified map of current data analytics for disaster response practice.

the dune using LiDAR Digital Elevation Map (DEM) data or identifying the disaster-prone area using twitter data, etc. The final step is decision-making, in which experts identify and choose alternative response operations including mobilizing resources or planning search and rescue operations, etc. Notably, in practice, the connections between these three processes are one-directional: the data acquisition step collects data and pushes them to the data processing step, and the data processing step delivers processing results to the decision-making step. In other words, there is no formal feedback loop in the system. The lack of feedback mechanisms causes the decision-makers to have little control over the specific tasks to be processed and the corresponding time requirement. In normal situations, the lack of feedback mechanism between these steps can be compensated by performing data collection, data processing, and decision-making in an iterative fashion until all information is obtained. Waiting for all information to be ready is highly infeasible in disastrous situations because of the necessity of making quick decisions in a dynamically changing environment. In contrast to normal situations, collection and processing of large geospatial datasets during extreme events require careful coordination and integration with decision-making processes. The overall challenges related to efficient use of large geospatial datasets during extreme events are summarized as below:

Challenge 1: Lack of clear understanding on the basic structure of big visual disaster data and their role in disaster management

Analysis of big visual disaster data offers tremendous opportunities in improving our understanding, modeling, and prediction of the impacts of coastal hazards on communities and ecosystems. While the big visual disaster data has widely adopted in routine coastal resilience applications, the role of this data in disaster management is controversial. It is widely agreed that big visual disaster data contains indispensable disaster information that can be integrated into ongoing decision processes, however, it is arguable whether such data is central or peripheral because the unclear structure of big visual disaster data cast doubt in the effective and efficient interpretation. To fully exploit the merit of big visual disaster data, it is necessary to revisit and characterize and the structure of the big spatial data in disaster situation awareness to make information available on time and at relevant level of decision-making in disaster management.

Challenge 2: Lack of understanding on the quality of big visual disaster data

The quality of big visual disaster data is rife with uncertainty (Fisher 1999). This uncertainty in data quality not merely refers to data accuracy (or error), but also includes other characteristics such as lineage, the goodness of fit for designated applications, etc. Data quality is always an issue in big visual disaster data related disaster response applications. Therefore, handling big visual disaster data requires the proper accommodation of uncertainty in data quality. Poorly handling of uncertainty, at best, result in inaccuracy of the information, and at worse, result in fatal errors. Awareness of data quality is principal for both data processing and decision-making. In data processing, it requires data quality analysis to provide prior knowledge of whether the processing results are right-fit to use or just a waste of time. On the other hand, in decision-making, the data quality determines how much trust decision-makers or experts can place in the information, and consequently determine the merits of the information. To this end, performing a comprehensive data quality analysis is equally important as processing big visual disaster data.

Challenge 3: Lack of formal modeling of processing goals, computational workflows in a distributed computing environment, and the coordination of decision-making and computational workflow

One of the significant challenges for using big visual disaster data in coastal resilience application is coordination of decision-making and computational workflow. This coordination requires a closing loop between decision-making and data processing. From the experts or decision-making perspective, it needs insights: key signals and tightly packaged summaries of relevant, intriguing disaster information. On the other hand, from the data processing perspective, it urges a clarified, well-defined goal, which they can convert to a series of feasible computation tasks. Currently, there is a huge shaded area between this decision-making and data processing: there is a lack of formal modeling of processing goals, computational workflows in a distributed computing environment, and the coordination of decision-making and computational workflow. Understanding of both technology and vernacular of decision-making is difficult. Mapping technology capabilities to vernacular of decision-making goals are even more complicated.

Challenge 4: Adaptive processing in time-sensitive applications

Disaster response involves making difficult decisions within a short time window. This determines that information extraction from big visual disaster is time sensitive in the same manner (Lippitt et al. 2014). The importance of getting timely information during natural disasters has recently motivated a nationwide survey of many U.S. emergency response organizations to understand the relationship between value and lag-time of the information in disaster response (Hodgson et al. 2014). In disasters, the merit of big visual disaster data diminishes rapidly as time goes on. Different from the routine processing that emphasis on maximizing performance (e.g., accuracy), disaster response applications allow a sacrifice of performance in trading for speed to meet the strict time budget. To this end, anticipating adaptive mechanism that could adjust the processing to the time budget remains has profound meaning.

3.3 A HURRICANE SANDY INSPIRED BIG DATA FRAMEWORK FOR COASTAL RESILIENCE INVESTIGATIONS WITH HETEROGENEOUS SPATIAL DATA

3.3.1 Geospatial Response to Hurricane Sandy

Driven by the abovementioned three primary needs, this study presents a Hurricane Sandy inspired Big Data Framework for Coastal Resilience Investigations with heterogeneous spatial data. Like during many other extreme events, geospatial products and tools are an essential part of every stage of disaster management during Hurricane Sandy, from planning through response, and recovery to mitigation of future events. However, unlike many other extreme events where the available spatial data are often limited in size and type, Hurricane Sandy has seen a surge of massive spatial data sets.

Figure 3.9 shows the type of spatial disaster data, either collected or identified during Hurricane Sandy. These datasets are imagery and point cloud data and can be more broadly defined as low-dimensional, spatio-temporal datasets, in which data elements are defined at points in a 2D/3D spatial coordinate system and over time.

The specific data sets considered in this study include various airborne LiDAR data sets collected at different points of time before and after the landing of Hurricane Sandy (Figure 3.9). First, airborne LiDAR data dated back to 2010 exist for the most of the New York-New Jersey metropolitan area and are archived in data repositories including Digital Coast and USGS Click. Second, on October 29, 2016, the day before Hurricane Sandy landed in New Jersey, the USGS Coastal and Marine Geology Program collected airborne LiDAR data along the New Jersey Coast using its Experimental Advanced Airborne Research LiDAR-B (EAARL-B) system. Immediately after the landing of Hurricane Sandy, NOAA collected airborne imagery followed by USGS EARL-B airborne LiDAR data collection. During the period of November 11–24, 2012, USACE conducted another wave of airborne LiDAR data collection along the New Jersey and New York coastal line. During the period of December 5–9, 2012, Rutgers conducted mobile LiDAR scanning of severely impacted coastal communities in the state of New Jersey and New York City. Throughout the disaster response period, street-level images of storm damage have also been collected by various damage assessment teams and citizens. Some of them were distributed through social media channels such as Facebook and Twitter. During the disaster recovery stage, more geospatial data sets have also been collected to assess recovery progress and future vulnerability. These data sets include 2014 USGS airborne LiDAR data collection along the coastal lines in the northeast region and mobile LiDAR data

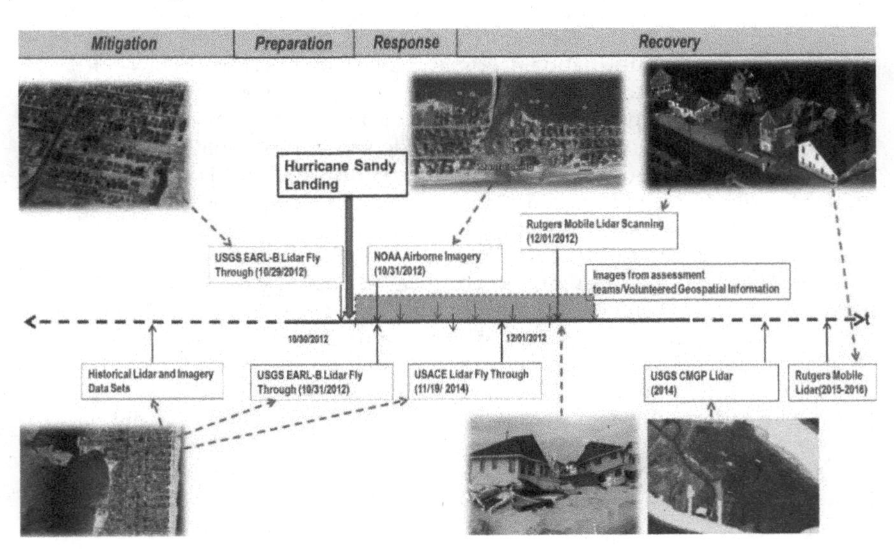

Figure 3.9 Hurricane Sandy-related 3D disaster.

collection in Ocean County, New Jersey in 2016. Collectively, these data sets are too massive to be efficiently managed and processed to derive scientific insights into ways of improving coastal resilience. In the following, this study characterizes the basic anatomy of big spatial disaster data to highlight the big data challenge in using these data sets in coastal resilience applications.

3.3.2 Data Analytic Framework

Existing analytical frameworks for interpreting 3D disaster datasets are insufficient for time-sensitive applications. For example, immediately after the landing of hurricane events, there is a great urgency to process the extensive and heterogeneous point cloud data, dynamically evolving in time as more data come in, and make sound decisions given limited and sparse resources. In this kind of scenarios, emergency response organizations would seek near real-time information about the extent of damages to homes and critical infrastructure systems such as transportation network, healthcare facilities, energy infrastructures, and wastewater treatment facilities. Data analysts in support of these organizations would construct a workflow of analytics steps consisting of selecting available point cloud data sources, querying point cloud data in geographic regions of interest by defining customized boundaries or overlaying vector data describing the locations of existing facilities, fusing point cloud data from different sources, selecting damage assessment analysis methods such as change detection with pre- and post-event data or classification of facility damages based on mono-event data, and aggregating extracted damage information into inputs to various decision planning models and tools that support search and rescue operations and restoration planning for critical infrastructures. They would also monitor information from social media outlets and potentially use them to prioritize data processing in specific regions. These types of analyses require efficiently querying of giant point cloud datasets, co-registration of heterogeneous point cloud data from different sensors on the fly, and near real-time detection and classification of damages to infrastructure systems and buildings.

Current ways of point cloud data management and analysis also pay little attention to the coordinated use of many interrelated analysis pipelines. For various point cloud data consumers, choices of analytic workflows have to factor in time, stakeholder information needs, and tolerance of errors. In large urban communities, people and infrastructure systems are often highly concentrated to achieve productive proximity. In many post-disaster scenarios, infrastructure stakeholders may have overlapping, or interdependent information needs due to the geographic interdependency of infrastructure systems. For example, to reduce the risk of secondary catastrophic events, natural gas pipeline operators would seek information on where the debris and overwash is deposited as this will impact their accessibility to critical gas shutoff valves (Zhou et al. 2016). At the same time, this information is also critically sought by other organizations like FEMA and local emergency response organizations. Therefore, it is clear that analytics applications should not be designed in a silo. Instead, there is a great need for mechanisms to design flexible and concerted geo-data workflows to maximize their utilities. Unfortunately, there appears to be no framework existing for tailoring and optimizing spatial analytics to needed capabilities in various phases in disaster management.

In addressing the abovementioned research gap, we proposed a data analytic framework (Figure 3.10) for the timely applications. In addition to the tradition data analytic framework described in Figure 3.8, we added time-sensitive features to the existing data analytic framework. In the following sections, we will describe the five essential elements in the data analytics framework.

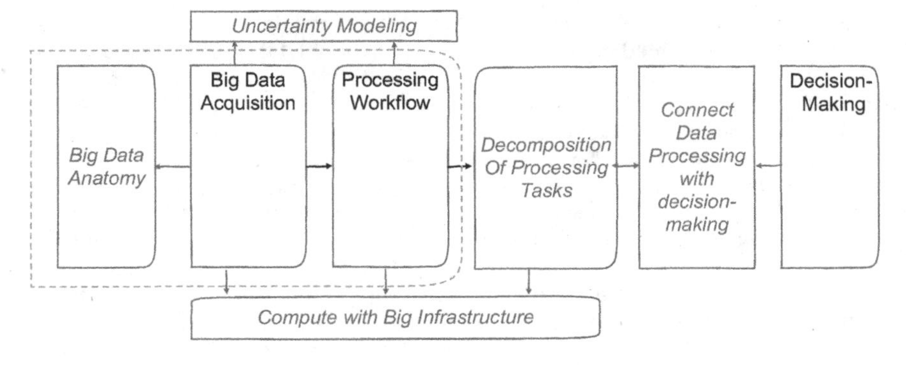

Figure 3.10 Data analytic framework.

3.3.3 Anatomy of Big Spatial Disaster Data

3.3.3.1 Volume

Although big data do not purely mean the large volume of data, data volume remains a major concern in disaster response and recovery missions, where how large the amount of data generated often determines what kind of protocols are used for storage and transferring and how much computation resources are required to process it. Advanced sensors such as mobile LiDAR systems can generate a large amount of data (Table 3.2) each day. For example, the Rutgers Mobile LiDAR survey system, each day (considering 8 hours of data collection), can generate around 43 gigabytes of LiDAR data in the format of pcap file and approximately 1 gigabyte of position file. Another 300 gigabytes of 360-degree imagery data would be captured along with LiDAR scanner. Thus, the sheer volume of the data generated each day from the mobile system alone is around 350 gigabytes. This study systematically analyzed the volume of the Hurricane Sandy related spatial data sets. Table 3.2. provides a quick summary of the volume of these data sets used in disaster response and recovery phases during Hurricane Sandy.

3.3.3.2 Data Structure

Data structure type refers to the structure or organization that data come with. In general, there are three primary data structure types including structured data,

Table 3.2: Volume of Hurricane Sandy-Related Spatial Disaster Datasets

Survey Type	Data Collection Date	Data Volume
Archived Airborne LiDAR	Archived	29.6GB
USGS EARL-B LiDAR	10/29/12	2.1GB
USGS EARL-B LiDAR	10/31/12	2.1GB
USACE LiDAR	11/19/14	21.2 GB
Rutgers Mobile LiDAR	12/01/12	575GB
USGS CMGP LiDAR	2014	105GB
Photos for SFM reconstruction	Streaming	20GB
Rutgers Mobile LiDAR	2015–2016	15TB

semi-structured data, and unstructured data. Structured data refers to the data that stored in pre-defined data models such as relational databases or spreadsheets. Most ancillary data are structured data, e.g., Excel spreadsheets, ArcGIS shapefile, etc. Unstructured data refer to data that do not have a pre-defined data model for information extraction. Most of the on-site remote sensing data are fallen into this category such as LiDAR data. Semi-structured data lies in between structured data and unstructured data. More specifically, it can be considered as a type of structured data, but lack a strict structure imposed by an underlying data model. One example is social media data. The text itself is structured data, but the time tag and location, as well as other information, add to the complexity. Given the uncertainty of disaster environment, situation awareness for disaster response requires the integration of data with different structured types. Table 3.3 depicts an example list of data required for disaster response (Figure 3.11).

3.3.3.3 Spatial Completeness

Beyond volume, variety, and variability, the challenge also arose from dealing with incompletion/inaccurate information. For example, one primary issue in using airborne LiDAR for building damage assessment is their vertical perspectives, which strictly limits their sensor readings to building roofs (Olsen 2013). In contrast, ground-based spatial sensing methods such as mobile LiDAR mostly capture data from the horizontal perspective and inevitably miss part of objects that are not visible from the driving paths. Figure 3.12 depicts a coverage analysis of roof and wall respectively using the Rutgers Mobile LiDAR system (MLS).

3.3.3.4 Veracity

The difference in data accuracy is also a common problem in conducting comparative analysis among various spatial data sets. For example, Table 3.4 lists the accuracy of various spatial disaster datasets expressed in terms of Residual Mean Square Error. The issue of accuracy also exists in single data sets. For example, this study analyzed to determine the consistency of vertical accuracy in mobile LiDAR datasets. The author compared the accuracy of two datasets: (1) USGS airborne LiDAR dataset; (2) Rutgers Mobile LiDAR dataset. The accuracy of USGS airborne LiDAR data is controlled via a network of ground control points. On the other hand, no control points were used in mobile LiDAR data collection. The accuracy of the mobile LiDAR data is calculated as the difference between the ground surface elevation detected using the airborne LiDAR and the ground surface elevation detected by mobile LiDAR. The comparison is conducted in various environmental conditions including urban environment (New Brunswick downtown), shoreline residential communities (Normandy Beach and Ortley Beach) and important infrastructures (Rockaway Bridge), which are representative for the dense population area. The analysis results show that mobile LiDAR performs better in open shoreline area (Table 3.5). The urban high-rise buildings, vegetation, traffic signals are potential obstructions that shade the GPS signals. Use of these data with different vertical accuracies in disaster response and recovery is a challenging endeavor as many analyses involve change detection between these data sets to detect damaged structures.

3.3.3.5 Velocity

Velocity is another challenge associated with big spatial disaster data. It does not only refer to how fast data are generated, but also refer to the need of speed in data analysis. Nowadays, spatial data can be collected at an almost unimaginable speed. For example, many MLSs can collect point measurement at 1 million points per second. More tremendous amount of data can be generated in the social

Table 3.3: An Example List of Data Required for Disaster Response

Type of Information	Datasets Required	Objective of Data	Data Structure Type	Data Repository Type	Data Access Type
Ancillary Data	Land use, boundary of administrative religion, roads	Baseline data for analysis	Structured data	In-house big data infrastructure storage/ Cloud-based big data Storage	Transactional, operational, and warehouse data
	Population	Baseline data for analysis	Structured data	In-house big data infrastructure storage/ Cloud-based big data storage	Transactional, operational, and warehouse data
Remote Sensing Data	Airborne LiDAR	Large area terrain mapping and rough scale damage identification	Unstructured data	In-house big data infrastructure storage	Transactional, operational, and warehouse data/ device or sensor access
	Mobile LiDAR	Detail damage assessment	Unstructured data	Disk-based big data storage/ In-house big data infrastructure storage	Device or sensor access
	Satellite Imagery	Long-time terrain or environment change simulation	Unstructured data	In-house big data infrastructure storage	Device or sensor access
	Oblique Imagery	Real-time disaster situation awareness	Unstructured data	Disk-based big data storage/ In-house big data infrastructure storage	Transactional, operational, and warehouse data/ device or sensor access
Crowdsourcing	Geo-tagged Imagery data	Real-time disaster situation awareness	Semi-structured/ Unstructured data	Cloud-based big data storage	Web or Internet access
	Geo-tagged message	Identify the heavily hit from victim report, understanding the sentiment of people from disaster-prone area	Semi-structured	Cloud-based big data storage	Web or Internet access
Emergency Service	Police department, Hospital, Fire Department, Public Broadcasting Service	Informing citizens disaster information and guide them in disaster response operations	Structured data	Disk-based big data storage	Web or Internet access/ device or sensor access

 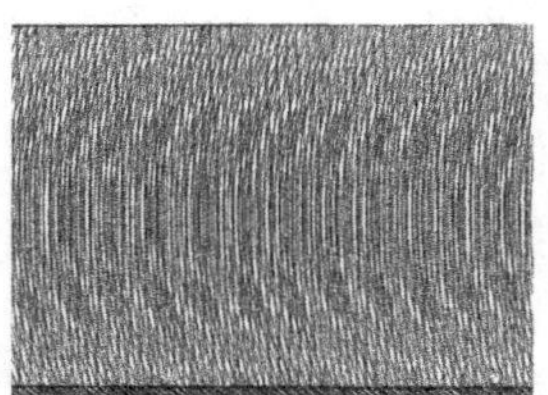 

Figure 3.11 Comparison of scanning pattern from different scanning platforms. (a) Static terrestrial LiDAR. (b) Mobile LiDAR system. (c) Airborne LiDAR system.

media: every minute, more than 100 hours of video are uploaded to YouTube, 5 thousand tweets are sent, and the type and the volume of this data are likely to increase. For major disasters, real-time or near real-time spatial data acquisition is critical and feasible. For example, USGS was able to collect the pre- and post-event data for the entire New Jersey and New York shoreline area within a day or two. During Hurricane Sandy, there is also a constant stream of social media based image data capturing the ever-dynamic disaster impact. While the importance of real-time data collection is well recognized during disastrous events, extracting meaningful information in a timely manner remains one of the most significant challenges.

3.3.4 Decomposition of Processing Tasks

While there appear to be a variety of ways to process the large volume of spatial data, the most common ones include generation of digital surface models, feature extraction, and change detection.

3.3.4.1 Digital Elevation Models

Disaster response can be impeded by the lack of precise terrain information. LiDAR and photogrammetry approaches have the capability to capture elevation data in large areas immediately, and the collected point cloud data can be used to create highly accurate 3D representations of the impacted terrain. This cartographic information plays a vital role in assessing damage, analyzing potential risks. The DEMs can be generated by eliminating the non-ground objects (Meng et al. 2009). A detailed review of different ground filtering algorithms can be found in Meng et al. (2010). Numerous studies have attempted to improve the performance of ground filtering algorithms by deploying approaches such as interpolation-based (Briese and Pfeifer 2001) or morphology-based methods (Kobler et al. 2007). However, these algorithms often require searching for neighbors, which are computationally expensive (Meng et al. 2010). The time sensitivity issue of the algorithms is only considered by few studies (Liu et al. 2007).

3.3.4.2 Feature Extraction

Object or feature extraction is much more complicated than generating DTMs (Mayer 2008). Prior to feature extraction, segmentation is the fundamental step for exploitation of 3D point clouds (Yang et al. 2015). One of the earliest studies in segmentation of point cloud data is conducted by Henderson and Bhanu (1982), who developed a region growing algorithm using a spatial proximity graph. Rabbani et al. (2006) introduced a widely used a k-nearest neighbors method for point cloud applications. However, it is not until Random Sample Consensus (RANSAC) (Schnabel et al. 2007) that the time efficiency is taken into

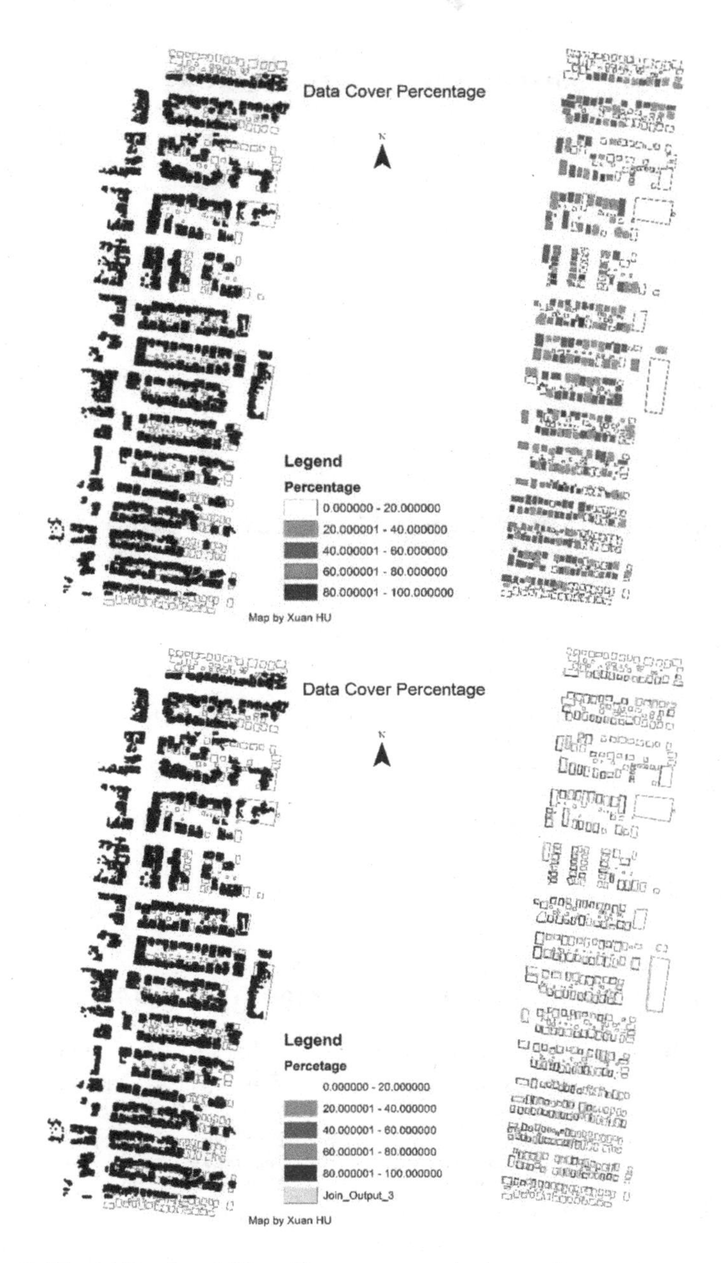

Figure 3.12 (a) Roof and (b) wall coverage analysis results.

serious consideration. In another attempt to reduce the computation cost, voxels-based segmentation algorithm is developed by Lim and Suter (2008). Distinctive features are further extracted by tilling with different parameters using fitting based on algorithms such as local fitting surfaces (Mongus et al. 2014) or support vector machines (Mountrakis et al. 2011). A list of different features extracted from LiDAR datasets is shown in Table 3.6.

Table 3.4: Resolution and Vertical Accuracy of Hurricane Sandy-Related 3D Disaster Datasets

Data Type	Resolution (pts/m2)	Vertical Accuracy (cm)
Archived airborne LiDAR	1–4	36.6
USGS EARL-B LiDAR	1–2	20
USGS EARL-B LiDAR	1–2	20
USACE LiDAR	1–4	8.2
Rutgers Mobile LiDAR	1000–8000	5
USGS CMGP LiDAR	1–4	6
SFM reconstruction (Zixiang et al. 2015)	500–2000	20
Rutgers Mobile LiDAR	2000–4000	5

Table 3.5: Accuracy of Mobile LiDAR Data in Different Environments

Environment	Locations	Airborne Data	Mean (m)	Standard Deviation (m)
Urban and Residential	New Brunswick Downtown	USGS NJ	−0.073	0.051
Urban and Schools	Rutgers Busch Campus	USGS NJ	−0.123	0.021
Shoreline and Residential	Normandy Beach	USGS NJ	−0.069	0.037
Shoreline and Residential	Ortley Beach	USGS NJ	−0.103	0.020
Urban and Infrastructure	Rockaway Bridge	USGS NY	0.023	0.044

Table 3.6: Different Features Extracted from LiDAR

Feature Category	Feature type	Reference
Building	3D Models	Rottensteiner and Briese (2002); Verma et al. (2006)
	Planar (Building footprint/Roof Polygons)	Rottensteiner et al. (2005); Awrangjeb et al. (2013) Henn et al. (2013)
Planimetrics	Roadways	Peterson et al. (2008); Olsen (2013)
Infrastructure	Transmission Lines	McLaughlin (2006); Jwa et al. (2009)
	Pipelines	Son et al. (2014)
	Storage Tanks	Fernández-Lozano et al. (2015)
Street Furniture	Street Light	Yu et al. (2015)
	Power/Telco Pole	Jwa et al. (2009)
	Fire Hydrant	Korah et al. (2011)
	Debris	Labiak et al. (2011)
Hydrologic Features	Channel Network	Passalacqua et al. (2010)

3.3.4.3 Change Detection

Change detection refers to the process of identifying meaningful difference over different observations (Singh 1989). As a result, change detection often requires at least two datasets with overlaps. Among the ten common change detection techniques described in Singh (1989), image differencing method is the most widely used one for LiDAR applications (Trinder and Salah 2012). In this method, two datasets are spatially registered, subtracted, and then pass through a user-specified threshold so that the significance can be identified. This procedure has frequently been employed in disaster situations such as landslide (Hsiao et al. 2004), earthquake (Zhang et al. 2006), and hurricane (Hatzikyriakou et al. 2015).

In this research, major processing tasks are decomposed into different operations. We will target core operation categories employed in applications that synthesize information from spatio-temporal sensor data in our research. The core operations produce different levels of data products that can be consumed by client applications. For example, a client application may request only a subset of satellite imagery data covering the east coast of the United States. The operations can also be chained to form analysis workflows to create other types of data products. An example workflow could be a pipeline of: [data cleaning → subsampling mapping → object segmentation → object classification → change detection] operations. Each operation's data access and processing patterns, as well as the composition of the analytics application, are important factors in I/O, communication, memory, and processing overheads. The data access and processing patterns range from local and regular, to indexed data access, to irregular and global access to data – please see the third column in Table 3.7. Local data access patterns correspond to accesses to a single data element or data elements within a small neighborhood in a spatial and temporal region (e.g., data cleaning and low-level transformations). Regular access patterns involve sweeps over data elements, while irregular accesses may involve accesses to data elements in a random manner (e.g., certain types of object classification algorithms, morphological reconstruction operations in object segmentation). Some data access patterns may involve generalized reductions and comparisons (e.g., aggregation) and indexed access (e.g., queries for data subsetting and change quantification).

3.3.4.4 Core Operation Categories

The composition of the analytics applications encapsulates several application-level data processing structures as well. First, original datasets can often be partitioned into tiles or chunks, and several categories of operations in Table 3.7 can be executed on each chunk independently. Spilting large datasets into chunks leads to a bag-of-tasks processing pattern. Second, processing of a single chunk or a group of chunks can be expressed as a hierarchical coarse-grain data flow pattern (Beynon et al. 2001; Plale and Schwan 2000; Tan et al. 2010). For example, transformation, filtering, mapping, and segmentation operations can be composed as a workflow. The segmentation operation itself may consist of a pipeline of lower level operations as well. Third, several types of operations such as aggregation and classification can be represented as MapReduce style (Dean and Ghemawat 2008; Dean and Ghemawat 2010) computations. The detail descriptions of each core operation categories are summarized as in Table 3.7.

Table 3.8 lists examples of application-specific operations as mapped to the core operation categories defined in Table 3.7. All three applications have similar operations, although they use spatio-temporal datasets for different purposes or may handle different data types. Thus, we argue that an efficient framework that can support the core operation categories can benefit a wide range of applications. Table 3.9 depicts Corresponding tools for the seven core operation

Table 3.7: Core Operation Categories in Spatial Disaster Data Processing

Core Operation Categories	Example Operations	Data Assess Pattern	Computation Complexity
Data Cleaning & Quality Control	Transformations to reduce effects of sensor/ measurement artifacts. Transform sensor acquired measurements to domain-specific variables	Mixture of local and global pattern	Moderate computational complexity
Low-Level Transformations	Transformations of a dataset to another format. E.g., coordinate transformation (such as UTM to GCS), value conversion (such as. RGB to grayscale conversion), or as geometry transformation (3D to 2D projection)	Mainly local pattern	Low to moderate, mainly data-intensive computations
Data Subsetting, Filtering, Subsampling	Select portions of a dataset corresponding to regions in the atlas and/or time intervals. Select portions of a dataset based on value ranges. Subsample data to reduce resolution and data size	Local as well as indexed pattern	Low to moderate, mainly data-intensive computations
Spatio-temporal Mapping & Registration	Create composite dataset from multiple spatially co-incident datasets. Create derived dataset from spatially co-incident datasets obtained at different times	Irregular local and global data pattern	Moderate to high computational complexity
Object Segmentation	Segment "base level" objects such as ground, road, dune, vegetation, and buildings. Extract features from "base level" objects	Irregular, but primarily local data pattern	High computational complexity
Object Classification	Classify "base level" individual objects at finer details such as utility poles, building types, and transportation assets through a possibly iterative combination of clustering, machine learning and human input (active learning)	Irregular local and global data patterns	High computational complexity
Change Detection, Comparison, and Quantification	Quantify changes over time in domain-specific low-level variables, base level objects, and high-level objects. Construct "change objects" to describe changes in low-level domain-specific variables, base level, and high-level objects. Spatial queries for selecting and comparing segmented regions and objects	Mixture of local and global data patterns as well as indexed	High Complexity and data-intensive computations

Table 3.8: Examples of Operations from the Example Application Scenarios Mapped to the Core Operation Categories

Operation Category	Weather Prediction	Monitoring and Change Analysis	Pathology Image Analysis
Data Cleaning and Low-Level Transformations	Remove anomalous measurements from MODIS and convert spectral intensities to the value of interest	Remove unusual readings. Convert signal intensities to color and other values of interest	Color normalization. Thresholding of pixel and regional grayscale values
Data Cleaning and Low-Level Transformations	Spatial selection/crossmatch to find the portion of a dataset that is corresponding to a given geographic region	Spatial selection/crossmatch to find portion of a dataset corresponding to a given geographic region	Selection of regions within an image. Thresholding of pixel values
Data Cleaning and Low-Level Transformations	Mapping tiles to map projection. Generation of a mosaic of tiles to get complete coverage	Registering low and high-resolution images corresponding to same regions	Deformable registration of images to an anatomical atlas
Object Segmentation	Segmentation of regions with similar land-surface temperature	Segmentation of buildings, trees, plants, etc.	Segmentation of nuclei and cells. Compute texture and shape features
Object Classification	Classification of segmented regions	Classification of buildings, trees, plants	K-means clustering of nuclei into categories
Spatio-temporal Aggregation	Time-series calculations on changing land and air conditions	Aggregation of labeled buildings, trees, plants into residential, industrial, vegetation areas	Aggregation of object features for per image features
Change Detection, Comparison, and Quantification	Spatial and temporal queries on classified regions and aggregation to look for changing weather patterns	Characterize vegetation changes over time and are	Spatial queries to compare segmented nuclei and features

Table 3.9: Corresponding Tools for the Seven Core Operation Categories

Core Operation Categories	Tools in Lastool	Tools in Cloud Compare	Tools in Terrasolid
Data Cleaning & Quality Control	LasControl, LasDuplicate, LasInfo, LasNoise, LasPrecisiom, LasReturn, LasThin, LasValidate, LasView	Noise Filter, SOR (Statistical Outlier Removal) filter, Remove Duplicate Points, Hidden Points Removal	TerraMatch: Calibration and Strip Adjustment, Tie Lines tools, Match tools (e.g., apply correction, find intensity correction), etc.
Low-Level Transformations	Blast2Dem, Blast2Iso, Las2Dem, Las2Iso, Las2Las, Las2Shp, Las2tin, las2Txt, Las2Zip, Shp2Las, LasPublish	Fit Tool (plane, sphere, 2D polygon, 2.5D quadric), Unroll, Rasterize and Contour Plot, Contour Plot to Mesh	Projection tool (Coordinate Transformations, Geoid adjustment), Convert Storage Format (to kmz, dgn, etc.)
Data Subsetting, Filtering, Subsampling	Las2Las, LasCanopy, LasClip, LasGrid, LasIndex, LasCoverage, LaSort, LasSpilt	Subsampling Tool (by random, space, octree)	Point Filtering Tools (by classification, intensity)
Spatio-temporal Mapping & Registration	LasColor, LasTrack, LasPlane	Align (point pairs picking), Match Boundary Box Centers, Match Scales, Fine Registration	TerraPhoto: Camera Calibration Tool, Color Correction Tool, Improving Image Positioning Tool, Color Points, and Selection Shapes Tools, Manage Trajectories Tool
Object Segmentation	Las2Boundry, LasClassify, LasHeight, LasGround	Label Connect Component, Cross Section/ Unfold, Section, Facet Detection, RANSAC shape detection	TerraScan: Macro Classification tool (Classify / By intensity; Classify / Surface Points, Classify Using Brush, etc.), Power Lines using Least Squares Fitting, TerraModel (Surface Modeling)
Object Classification	LasClassify, LasHeight, LasGround	CANUPO Classification, Cloth Simulation Filter (CSF)	TerraScan: Macro Classification tool (Classify / By intensity; Classify / Surface points, Classify Using Brush, etc.)
Change Detection, Comparison, and Quantification	-	Compute 2.5D Volume	TerraScan Change Detection Tool

categories in three major cloud computing softwares: Lastool, Cloud Compare, and Terrasolid respectively.

3.3.5 Identify the Uncertainty Associated with Big Data Acquisition and Processing

While spatial information (e.g., LiDAR, high-resolution imagery, etc.) facilities the rapid collecting of spatial disaster data for situation awareness, the uncertainty associated with the spatial data could be a defect that devalues the merit of the spatial data. Poor handling of the uncertainty can, at best, lead to errors in the knowledge of what the data represents and at worst can have fatal consequences (Fisher 1999). Therefore, without a clear understanding of the uncertainty in spatial data, the value of the big spatial data will be discounted as experts and decision-makers may question the accuracy of spatial data and especially their derived products using fast computing techniques and even be reluctant to use it for their judgments (Gahegan and Ehlers 2000). In general, there are two types of uncertainties (Table 3.10).

Raw data uncertainty refers to the inherent uncertainty of the raw datasets. The objective of studying raw data uncertainty is to identify the uncertainty in data acquisition and derive methods for data quality control. Hasselman et al. (2005) defined the studying of raw data uncertainty as the research question of "How accurate is the data perfect representation of the real world?" There are numerous researches on raw data uncertainty often in the form of uncertainty (Beard et al. 1991; Fisher 1999), error (Goodchild 1994), accuracy (Goodchild 1994), data quality (ISO 2002), etc. Known factors contributing error in LiDAR data include, but are not limited to, errors in LiDAR systems, navigation systems, and system calibration. Mezian et al. (2016) studied the impact factors for system accuracy on two aspects: laser scanner accuracy and navigation system accuracy. The accuracy of the laser scanning accuracy is determined by properties of targeted objects (roughness, reflectivity), scanner mechanism precision (mirror center offset) as well as weather conditions (temperature, humidity) (Mezian et al. 2016; Soudarissanane et al. 2008).

The other type of uncertainty is uncertainty trade-off, which is focused on leveraging between data characteristic and algorithm performance (time, accuracy). Hasselman et al. (2005) defined the study of uncertainty trade-off as "Is the data accurate enough for the specific application?" In real-world applications, it is essential to have a comprehensive understanding on uncertainty trade-off. Understanding the uncertainty trade-off is crucial in disaster environments where time is limited and computational resources are often constrained. Disaster applications desire not only computational speed but also acceptable uncertainty level. Without properly addressing either the timely or the uncertainty issues, the value of the LiDAR data-derived products will be demerit as experts and decision-makers may cast doubt and even reluctant to use them for their judgments (Gahegan and Ehlers 2000). In the context of LiDAR data-derived DEMs, larger interpolation cell size results in DEMs with smaller file sizes and less computation time. Meanwhile, the larger interpolation cell size also potentially increases the information loss while generating these DEMs. Achieving an ideal trade-off between computation time and uncertainty level requires a "frugal" interpolation cell size. In literature, most of the studies focus on the uncertainties of different interpolation methods. A comprehensive study comparing errors caused by different interpolation methods can be found in Bater and Coops (2009). However, very little work can be found on how interpolation cell size is related to the information loss in generating DEMs with LiDAR data. To this end, this study presented the studying of uncertainty trade-off as an essential element in the data analytics framework.

Table 3.10: Summary of Different Uncertainty Analysis Needs in Different Phases

Type	Data Acquisition Phase	Application Phase
Uncertainty Type	Raw Data uncertainty	Uncertainty Trade-off
Objective	Identify the uncertainty in data acquisition and derive method for data quality control	Leverage between data characteristic and algorithm performance (time, accuracy)
Research question	How accurate is the data prefect representation of the real world? (Hasselman et al. 2005)	Is the data accurate enough for the specific application? (Hasselman et al. 2005)
Methods	Data accuracy/error analysis	Sensitivity analysis/ Uncertainty (error) modeling
Uncertainty Representation	Value (Root Mean Square Error, Standard deviation)	Trade-off

3.3.6 Computing with Big Data Infrastructure

With the time requirement in disaster response, computing infrastructure systems are undergoing fundamental transformations. The first transformation is the shift from CPU computing to Graphics Processing Unit (GPU) computing (Liu 2013). CPU and GPU have rudimentary different design architectures. A CPU is designed for sequential serial processing with sophisticated control logic. Architecturally, a CPU has only a few cores that can handle a limited amount of tasks simultaneously with lots of cache memory. In contrast, a GPU is designed for massively parallel processing with simple control logic. The ability of a GPU with 100+ cores to process thousands of threads can scale of the processing ability by 100 times over a CPU alone. Therefore, GPUs are especially well-suited for arithmetic parallel tasks. GPU-accelerated computing has now grown into a mainstream movement (Owens et al. 2008). One the one hand, hardware manufacturers such as NVIDIA and AMD, have been transforming GPUs into a computational powerhouse. On the other hand, GPU computing is supported by prevailing operation systems such as Apple (with OpenCL) and Microsoft (using DirectCompute). Concurrently, the exploding GPU capability has attracted more and more researchers to use it to cope with big spatial data. For instance, Yuan (2012) examined the performance of CUDA-enabled GPU for LiDAR processing. The author concluded that the proposed GPU method could scale up a 30-fold speed increase over a similar sequential algorithm. Similar scalability can be found in (Lukač and Žalik 2013) who developed a GPU-based roofs' solar potential estimation algorithm. More systematically, Plaza and Chang (2007) investigated insights as well as the challenge of GPU-based time-Critical Remote Sensing Applications.

The second transformation is the shift from processing batch data to streaming data (or synonym for real-time and near real-time data) and interactive analysis. Batch processing is designed for "data at rest" as a result, might have "medium to high latency" (or a response time from seconds to a few hours). MapReduce is a typical framework for batch processing. The Apache Hadoop framework enables the execution of applications on large computer cluster systems through the implementation of the MapReduce computational paradigm. Because of the "medium to high latency," stream processing comes into being to satisfy fast data needs. Wähner (2014) stated that stream processing is most suitable for processing streaming sensor data. Typical computing framework for stream processing includes Apache Spark and Apache Storm. Another trend for big data analytics is interactive analysis. Interactive analysis or sometimes refers to "human in the loop," is a set of techniques that combining computation power of machines and with the perceptive and cognitive capabilities of humans, in order to extract knowledge from large and complex datasets. The anticipation of human interaction can be more effective in dealing with unscheduled tasks and unpredictable disturbance. Moreover, it will go beyond the bottleneck of fully automation algorithms. Typical interactive analysis tools include Google's Dremel, Apache Drill, etc. A summary of the big data analytic tools is shown in Table 3.11.

Table 3.11: A Catalog of Big Data Analytic Tools

Category	Tools	Application Area
Batch processing tools	Apache Hadoop & MapReduce	Data-intensive distributed applications
	Dryad	Parallel and distributed processing of large datasets using a very small cluster or a large cluster
	Apache Mahout	Large-scale data analysis applications with scalable and commercial machine learning techniques
	Apache Spark	Batch and stream processing of large datasets
	Jaspersoft BI Suite	Report generation from columnar databases
	Pentaho Business Analytics	Report generation from both structured and unstructured large volume of data
	Talend Open Studio	Visual analysis of big data sets
Real-time stream processing tools	Storm	A distributed and fault-tolerant real-time computation system for processing limitless streaming data
	S4	A general-purpose, distributed, scalable, fault-tolerant, pluggable computing platform for processing continuously unbounded streams of data
	SQLstream s-Server	Processing of large-scale streaming data in real-time
	Splunk	A real-time and intelligent big data platform for exploiting information from machine-generated big data
	Apache Kafka	A high-throughput messaging system for managing streaming and operational data via in-memory analytical techniques for obtaining real-time decision-making
	SAP Hana	An in-memory analytics platform aimed at real-time analysis of business processes, predictive analysis, and sentiment data
Interactive analysis tool	Google's Dremel	A system for processing nested data and capable of running aggregation queries over trillion-row tables in seconds
	Apache Drill	A distributed interactive big data analysis tool capable of supporting different query languages, data formats, and data sources

The third transformation is anticipating cloud computing to cope with big disaster data. As the development of sophisticated sensors tends to generate more data, the capability to process this vast and complex data could easily go beyond the power of regular desktop computers (Dorband et al. 2003). As mentioned in previous paragraphs, GPU computing facilities the fast computing. However, there are some limitations for GPU computing in local computers such as small onboard texture memory. To boost the capability of GPU computing, one solution is to deploy high-performance clusters. Supercomputers, hosted by either universities or research institutes (e.g., Rutgers Caliburn) or commercial cloud computing companies (e.g., Amazon EC2, Microsoft Azure) can assist in reaping the benefits of GPU computing while avoiding its limitations (Fan et al. 2004). In traditional, High-Performance Computing (HPC) was the exclusive domain of government agencies. Yet, because cluster computing is proven to have the robust computational capability, relatively low-cost budget (Yang and Chen 2010), more and more agencies begin to implement HPC for their applications especially in dealing with massive data in the time-sensitive environment (Yang et al. 2011). Another method to handle the overwhelming spatial data is edge computing. In contrast to process data at central facilities that have advanced computing capabilities, edge computing is an optimized cloud computing system aiming at distributing the computation workloads from centralized points to the logical extremes of a network. The edge computing paradigm makes it feasible for the enormous amount of data that collected at the edge devices to be seamlessly processed at the edge as well in an efficient and timely manner. Moreover, state-of-the-art data-driven edge processing framework such as the one proposed by Renart et al. (2017), allows users to define data-driven reactive behaviors that can efficiently exploit data content and location to dynamically and autonomously decide the way data are processed. The emerging needs for fast processing big spatial data fostered the development of cloud computing, which is evolutional in accelerating the computations related with information extraction in remote sensing (Lee et al. 2011).

3.3.7 Connecting Data Processing with Decision-Making Models

Disaster response involves humanitarian tasks that require a comprehensive understanding of disaster situation. These humanitarian tasks cannot be accomplished by either decision-makers or data processing teams alone. Ideally, decision-makers need to be fully aware of what information can be processed and what kind of information cannot be processed within specific time constraints so that they could determine the optimal job sequence. However, decision-makers or experts might not necessarily have the domain knowledge in data processing; and there exists no cognition model in connecting the data processing tasks and decision-making processes. Notably, for large image datasets, it is difficult to provide an insight of what the information details and uncertainty would be like prior to the data processing. On the other hand, without specification of processing tasks form decision-makers, data processing teams may generate abundant or even worthless information, resulting in wasting of time and computational resource. In sum, an efficient response requires actions of decision-makers, data processing teams as well as collaboration between them. Without losing generify, primary complexes of the collaboration workflow is shown in Figure 3.13.

First, experts or decision-makers from different agencies need to settle down a list of core processing goals and hand it to the data processing teams. Disaster decision-making is a complex system that involves different level information. Estimating the cascading effect of a disaster does not only need information

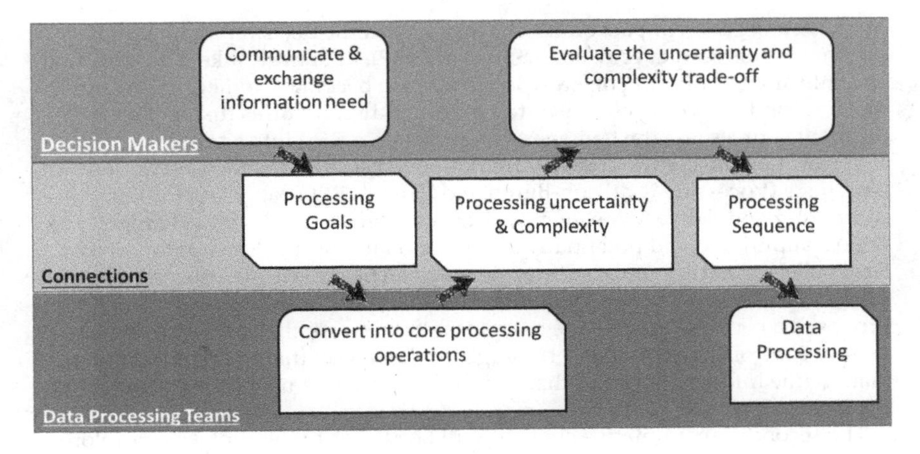

Figure 3.13 Connecting data processing with decision-making models.

on the destruction of critical infrastructures (e.g., pipeline, power line, dunes, etc.) and communications lines, but also intelligence on social, organizational and, economic structures that support the normal functioning of a community (Comfort et al. 2004). There could be conflicting roles structures in the experts and decision-makers (Bharosa et al. 2010). Therefore, it is critical to identify what kind of information are essential for response operations.

Second, data processing teams need to present the uncertainty and complexity of each processing goals to the decision-makers in a way that decision-makers even without domain knowledge can easily understand. One of the significant barriers for experts for job scheduling during a response is lacking sufficient knowledge in the uncertainty and processing complexity. In addressing this issue, the abstract processing goals are broken down into the core operation categories described in Table 3.7. In each operation categories, the uncertainty and complexity are provided as an index system, which enables experts to have a taste of what are the uncertainty and complexity associated with each processing tasks.

Finally, decision-makers need to send feedback to the data processing teams on the data processing sequence. The vast volume of available during a disaster often overwhelms the processing capability of computation infrastructure resulting in information overload. On the other hand, information for disaster response is time-critical (Horan and Schooley 2007). The merit of information in humanitarian relief is not only determined by the content of the information, but also by the time value of the information. To achieve an effective and efficient response, the data processing need to fulfill the data processing tasks according to the information demand orders. (Starcke and Brand 2012). Under such circumstance, it is indispensable for decision-makers to establish a chronological sequence of data processing jobs (Bharosa et al. 2010).

3.3.8 Future Improvement

Two transformative changes will likely occur in big spatial disaster data applications. The power of crowdsourced data may be an asset that could lead to substantial discoveries during extreme events. However, uses of crowdsourcing data during extreme events still face considerable skepticism. Many challenging research questions remain to be addressed. The spread of noise or fake data can have a negative impact on the individual and the decision-making during extreme events (Gupta et al. 2013; Shu et al. 2017). Therefore, fake data detection and information filtering on crowdsourced data become an emerging research topic. Second, handling crowdsourcing information requires fusing data from different sources (e.g., text, images, 3D spatial data, etc.) into a standard "readable" format. Most of the crowdsourcing-related studies are focused on Natural Language Processing (Callison-Burch and Dredze 2010; Sabou et al. 2012). The computing and delivery cost of crowdsourced data remain relatively high. Edge computing could potentially be a solution to lower the computation cost. Data collected at the emulated edge devices can be seamlessly processed at the edge in an efficient and timely manner. More importantly, Renart et al. (2017) proposed a content-driven edge computing framework that crowdsourced data quality control is performed in the edge level in a way that only critical information (the information rules that satisfied the rules by users) are submitted to the core.

The second transformative changes will be related to the explosive development of artificial intelligence approaches and algorithms, which are foundational blocks in the atomic operations including object segmentation and object recognition. The recent Alpha Go and Alpha Go Zero have proven the power of the emerging machine learning algorithms, either supervised or unsupervised, in solving the complicated and realistic problems. There are numerous attempts in deploying these machine learning algorithms to handle spatial data such as LiDAR data (Gleason and Im 2012), imagery (Marjanović et al. 2011), and text (Gupta et al. 2013). Convolutional neural networks for 3D point cloud classification have been extensively studied recently owing much to the rising of driverless cars. Nevertheless, point clouds collected from ranging devices on driverless cars have very different resolution and patterns when compared with point clouds collected from disaster impact mapping systems. Establishing a well-annotated extreme event related geospatial data sets appears to be a fundamental task to be done. Considering the fact that more and more geospatial data have been collected in recent extreme events, it is reasonable to expect that deep learning techniques will become prominent tools in information extraction from big spatial disaster data.

3.4 CONCLUSION

Severe weather events such as hurricanes, ice storms, surge, and flooding have been occurring across the United States and around the world, threatening places where economic and industrial activities are heavily concentrated. In the face of these natural disasters, building community resilience is essential to reduce the loss of livelihoods, economic cost, and social disruption. These extreme events are now increasing observed and monitored with a loosely coupled network of geospatial sensors. For instance, in recent years, because state and federal agencies have made airborne LiDAR data collection a priority, post-storm LiDAR collection is now routine after large surge event, and the vast amount of disaster data are now freely available online. In another example, emerging high-resolution sensing systems such as /mobile LiDAR have also been deployed for damage data collection during recent events such as Superstorm Sandy, generating an unprecedented amount of visual disaster data. Lastly, VGI, such as geo-tagged disaster photos, is a new breed of disaster data which methods have produced large and heterogeneous spatial disaster datasets spanning multiple spatial and temporal scale and with varying levels of confidence. Analysis of these datasets offers tremendous opportunities in improving the resilience and adaptability of cities in the face of future natural disasters.

Despite the high values in these data sets, the vast size and complex processing requirements of these new data sets make it challenging to efficiently use them in city management applications, in particular, emergency situations.

In the second section, we identified two significant purposes for big spatial disaster data: long-term "capability" building information and short-term "adaptability" enabling information. To cope with these two different purposes, we revisited the applications of four major types of spatial disaster data: (1) ancillary geospatial, (2) imagery data, (3) LiDAR data, and last (4) VGI. We then summarized the four challenges of using big spatial data for disaster response including. First, there is lacking clear understanding on the basic structure of big visual disaster data and their role in disaster management. Second, the knowledge on the quality and uncertainty associated with the big spatial disaster data remains insufficient. Third, there are few studies on formal modeling of processing goals, computational workflows in a distributed computing environment, and the coordination of decision-making and computational workflow. Finally, specifically for short-term "adaptability" enabling, it urges an adaptive process to adjust data processing to the time bounds requirement.

Driven by the growing needs of deploying spatial disaster data for more efficient response, in section three, we presented our research progress in designing data analytics frameworks during extreme events to integrate, share, and process these large data sets for an array of critical disaster management tasks. We first characterized the basic anatomy of big spatial data from six aspects: volume, variability, velocity, data structure, spatial completeness, and veracity. Moreover, to standardize the data processing, we targeted core operation categories employed in applications that synthesize information from spatio-temporal sensor data in our research. The core operations produce different levels of data products that can be consumed by client applications. Then we highlighted the uncertainty issues associated with both big data acquisition and application phase. A central component of our study is on how to use big data infrastructure to accelerate the processing of the massive amount of geospatial data, in particular streaming data, such that crucial insights can be extracted from the data within a realistic time-bound and time-sensitive decisions can be made to optimize city operations during extreme events. We revisited three undergoing fundamental transformations in computing. Another missing puzzle in the current using of big spatial disaster data is the collaboration between decision-makers and data processing team. We proposed a collaboration workflow to connect the data processing with the decision-making process.

Last but not least, the discussion section pointed two advanced issues in deploying big 3D data during extreme events. The first issue is to promote the using of crowdsourced data for extreme events. The second issues are transplanting the advanced machine learning algorithms for extreme events. We identified two critical needs for this issue including (1) establish well-annotated database, and (2) formulate the information needs.

REFERENCES

Allahbakhsh, M., Benatallah, B., Ignjatovic, A., Motahari-Nezhad, H. R., Bertino, E., and Dustdar, S. (2013). "Quality control in crowdsourcing systems: Issues and directions." *IEEE Internet Computing*, 17, 76–81.

Awrangjeb, M., Zhang, C., and Fraser, C. S. (2013). "Automatic extraction of building roofs using LIDAR data and multispectral imagery." *ISPRS Journal of Photogrammetry and Remote Sensing*, 83, 1–18.

Bater, C. W., and Coops, N. C. (2009). "Evaluating error associated with lidar-derived DEM interpolation." *Computers & Geosciences*, 35(2), 289–300.

Beall, C., Lawrence, B. J., Ila, V., and Dellaert, F. (2010). "3D reconstruction of underwater structures." *IEEE*, 4418–4423. In Intelligent Robots and Systems (IROS), 2010 IEEE/RSJ International Conference on (pp. 4418–4423). IEEE.

Beard, M. K., Buttenfield, B. P., and Clapham, S. B. (1991). NCGIA research initiative 7: Visualization of spatial data quality: Scientific report for the specialist meeting 8–12 June 1991, Castine, Maine, National Center for Geographic Information and Analysis.

Becker, J., Manville, V., Leonard, G., and Saunders, W. (2008). "Managing lahars the New Zealand way: A case study from Mount Ruapehu volcano." *Natural Hazards Observer*, 32, 4–6.

Beynon, M. D., Kurc, T., Catalyurek, U., Chang, C., Sussman, A., and Saltz, J. (2001). "Distributed processing of very large datasets with datacutter." *Parallel Computing*, 27(11), 1457–1478.

Bharosa, N., Lee, J., and Janssen, M. (2010). "Challenges and obstacles in sharing and coordinating information during multi-agency disaster response: Propositions from field exercises." *Information Systems Frontiers*, 12, 49–65.

Bhatla, A., Choe, S. Y., Fierro, O., and Leite, F. (2012). "Evaluation of accuracy of as-built 3D modeling from photos taken by handheld digital cameras." *Automation in Construction*, 28, 116–127.

Briese, C., and Pfeifer, N. (2001). "Airborne laser scanning and derivation of digital terrain models." In 5th Conference on Optical 3-D Measurement Techniques, edited by A. Gruen and H. Kahmen, 80–87. Vienna: Optical 3D Measurement Techniques V, Technical University.

Brilakis, I., Fathi, H., and Rashidi, A. (2011). "Progressive 3D reconstruction of infrastructure with videogrammetry." *Automation in Construction*, 20, 884–895.

Callison-Burch, C., and Dredze, M. (2010). "Creating speech and language data with Amazon's mechanical Turk," *Proceedings of the NAACL HLT 2010 Workshop on Creating Speech and Language Data with Amazon's Mechanical Turk*, Association for Computational Linguistics, 1–12.

Chen, F., Zhai, Z., and Madey, G. (2011). "Dynamic adaptive disaster simulation: Developing a predictive model of emergency behavior using cell phone and GIS data." In Proceedings of the 2011 workshop on agent-directed simulation, ADS '11, Boston, USA, 3–7 April 2011, pp. 5–12. *Society for Computer Simulation International.*

Comfort, L. K., Ko, K., and Zagorecki, A. (2004). "Coordination in rapidly evolving disaster response systems: The role of information." *American Behavioral Scientist*, 48, 295–313.

Csanyi, M. N., and Toth, C. K. (2007). "Improvement of lidar data accuracy using lidar-specific ground targets". Photogrammetric Engineering & Remote Sensing, 73(4), 385–396.

Dai, F., and Lu, M. (2010). "Assessing the accuracy of applying photogrammetry to take geometric measurements on building products." *Journal of Construction Engineering and Management*, 136, 242–250.

Dean, J., and Ghemawat, S. (2008). "Mapreduce: Simplified data processing on large clusters." *Communications of the ACM*, 51(1), 107–113.

Dean, J., and Ghemawat, S. (2010). "Mapreduce: A flexible data processing tool." *Communications of the ACM*, 53(1), 72–77.

Dorband, J., Palencia, J., and Ranawake, U. (2003). "Commodity computing clusters at Goddard Space Flight Center." *Journal of Space Communication*, 1, 113–123.

Earle, P. S., Bowden, D. C., and Guy, M. (2012). "Twitter earthquake detection: Earthquake monitoring in a social world." *Annals of Geophysics*, 54(6).

Ezequiel, C. A. F., Cua, M., Libatique, N. C., Tangonan, G. L., Alampay, R., Labuguen, R. T., Favila, C. M., Honrado, J. L. E., Canos, V., Devaney, C., et al. (2014). "UAV aerial imaging applications for post-disaster assessment, environmental management and infrastructure development." Proceedings of the 2014 International Conference on Unmanned Aircraft Systems (ICUAS) Orlando, Fl, 274–283, 2014.

Fan, Z., Qiu, F., Kaufman, A., and Yoakum-Stover, S. (2004). "GPU cluster for high performance computing." Proceedings of the 2004 ACM/IEEE conference on Supercomputing. IEEE Computer Society, Pittsburgh, PA, 47.

Fernández-Lozano, J., Gutiérrez-Alonso, G., and Fernández-Morán, M. Á. (2015). "Using airborne LiDAR sensing technology and aerial orthoimages to unravel roman water supply systems and gold works in NW Spain (Eria Valley, León)." *Journal of Archaeological Science*, 53, 356–373.

Fisher, P. F. (1999). "Models of uncertainty in spatial data." *Geographical Information Systems*, 1, 191–205.

Gahegan, M., and Ehlers, M. (2000). "A framework for the modelling of uncertainty between remote sensing and geographic information systems." *ISPRS Journal of Photogrammetry and Remote Sensing*, 55, 176–188.

Garson, G. D., and Biggs, R. S. (1992). *Analytic Mapping and Geographic Databases*, Issue 87. Thousand Oaks, CA: SAGE.

Gillespie, T., Chu, J., Frankenberg, E., and Thomas, D. (2007). "Assessment and prediction of natural hazards from satellite imagery." *Progress in Physical Geography*, 31(5), 459–470.

Gleason, C. J., and Im, J. (2012). "Forest biomass estimation from airborne lidar data using machine learning approaches." *Remote Sensing of Environment*, 125, 80–91.

Gong, J., Zhou, H., Gordon, C., and Jalayer, M. (2012). "Mobile terrestrial laser scanning for highway inventory data collection." *Computing in Civil Engineering* (2012), 545–552.

Goodchild, M. F. (1994). "Integrating GIS and remote sensing for vegetation analysis and modeling: Methodological issues." *Journal of Vegetation Science*, 5(5), 615–626.

Goodchild, M. F., and Glennon, J. A. (2010). "Crowdsourcing geographic information for disaster response: A research frontier." *International Journal of Digital Earth*, 3, 231–241.

Gupta, A., Lamba, H., Kumaraguru, P., and Joshi, A. (2013). "Faking sandy: Characterizing and identifying fake images on twitter during hurricane sandy." *ACM*, 729–736.

Han, Y., Sheth, A., and Bussler, C. (1998). "A taxonomy of adaptive workflow management." In Workshop of the 1998 ACM Conference on Computer Supported Cooperative Work, 1–11.

Hasselman, T., Wathugala, G., Urbina, A., and Paez, T. L. (2004) "Top-down vs. bottom-up uncertainty quantification for validation of a mechanical joint model." *Proceedings of 23rd International Modal Analysis Conference*, Orlando, FL.

Hatzikyriakou, A., Lin, N., Gong, J., Xian, S., Hu, X., and Kennedy, A. (2015). "Component-based vulnerability analysis for residential structures subjected to storm surge impact from Hurricane sandy." *Natural Hazards Review*, 05015005.

Henn, A., Gröger, G., Stroh, V., and Plümer, L. (2013). "Model driven reconstruction of roofs from sparse LIDAR point clouds." *ISPRS Journal of Photogrammetry and Remote Sensing*, 76, 17–29.

Hirokawa, R., Kubo, D., Suzuki, S., Meguro, J.-I., and Suzuki, T. (2007). "A small UAV for immediate hazard map generation." In AIAA Infotech@ Aerospace 2007 Conference and Exhibit, 2725.

Hodgson, M. E., Battersby, S. E., Davis, B. A., Liu, S., and Sulewski, L. (2014). "Geospatial data collection/use in disaster response: A United States nationwide survey of state agencies." *Cartography from pole to pole*, Springer, Berlin, 407–419.

Horan, T. A., and Schooley, B. L. (2007). "Time-critical information services." *Communications of the ACM*, 50(3), 73–78.

Hsiao, K., Liu, J., Yu, M., and Tseng, Y. (2004). "Change detection of landslide terrains using ground-based LiDAR data." *Proceedings of the XXth ISPRS Congress*, Istanbul, Turkey, Commission VII, WG, 5.

International Organization for Standardization (ISO). (2002). "Guidelines for quality and/or environmental management systems auditing." *The International Organization for Standardization*. ISO 19011: 2002-10.

Izadi, S., Kim, D., Hilliges, O., Molyneaux, D., Newcombe, R., Kohli, P., Shotton, J., Hodges, S., Freeman, D., Davison, A., et al. (2011). "KinectFusion: real-time 3D reconstruction and interaction using a moving depth camera." In Proceedings of the 24th annual ACM symposium on User interface software and technology, pp. 559–568. *ACM*, 2011.

Joyce, K. E., Wright, K. C., Samsonov, S. V., and Ambrosia, V. G. (2009). Remote sensing and the disaster management cycle, INTECH Open Access Publisher.

Jwa, Y., Sohn, G., and Kim, H. (2009). "Automatic 3d powerline reconstruction using airborne lidar data." *International Archives of the Photogrammetry Remote Sensing*, 38(Part 3), W8.

Kobler, A., Pfeifer, N., Ogrinc, P., Todorovski, L., Oštir, K., and Džeroski, S. (2007). "Repetitive interpolation: A robust algorithm for DTM generation from Aerial Laser Scanner Data in forested terrain." *Remote Sensing of Environment*, 108(1), 9–23.

Korah, T., Medasani, S., and Owechko, Y. (2011). "Strip histogram grid for efficient lidar segmentation from urban environments." *Proceedings of Computer Vision and Pattern Recognition Workshops (CVPRW), IEEE Computer Society Conference on, IEEE*, 74–81.

Labiak, R. C., Van Aardt, J. A., Bespalov, D., Eychner, D., Wirch, E., and Bischof, H.-P. "Automated method for detection and quantification of building damage and debris using post-disaster LiDAR data." *Proceedings of SPIE Defense, Security, and Sensing, International Society for Optics and Photonics*, 80370F-80370F-80378.

Lee, C. A., Gasster, S. D., Plaza, A., Chang, C.-I., and Huang, B. (2011). "Recent developments in high performance computing for remote sensing: A review." *IEEE Journal of Selected Topics in Applied Earth Observations and Remote Sensing*, 4, 508–527.

Lefsky, M. A., Cohen, W. B., Parker, G. G., and Harding, D. J. (2002). "Lidar remote sensing for ecosystem studies: Lidar, an emerging remote sensing technology that directly measures the three-dimensional distribution of plant canopies, can accurately estimate vegetation structural attributes and should be of particular interest to forest, landscape, and global ecologists." *BioScience*, 52, 19–30.

Lim, E. H., and Suter, D. (2008). "Multi-scale conditional random fields for over-segmented irregular 3d point clouds classification." *IEEE Computer Society Conference on Computer Vision and Pattern Recognition Workshops (CVPRW'08)*, IEEE, 1–7.

Lippitt, C. D., Stow, D. A., and Clarke, K. C. (2014). "On the nature of models for time-sensitive remote sensing." *International Journal of Remote Sensing*, 35, 6815–6841.

Liu, L. (2013). "Computing infrastructure for big data processing." *Frontiers of Computer Science*, 7, 165–170.

Liu, X., Zhang, Z., Peterson, J., and Chandra, S. (2007). "The effect of LiDAR data density on DEM accuracy." *Proceedings of the International Congress on Modelling and Simulation (MODSIM07)*, Christchurch, New Zealand.

Lukač, N., and Žalik, B. (2013). "GPU-based roofs' solar potential estimation using LiDAR data." *Computers & Geosciences*, 52, 34–41.

Marjanović, M., Kovačević, M., Bajat, B., and Voženílek, V. (2011). "Landslide susceptibility assessment using SVM machine learning algorithm." *Engineering Geology*, 123(3), 225–234.

Mayer, H. (2008). "Object extraction in photogrammetric computer vision." *ISPRS Journal of Photogrammetry and Remote Sensing*, 63(2), 213–222.

McLaughlin, R. A. (2006). "Extracting transmission lines from airborne LIDAR data." *IEEE Geoscience and Remote Sensing Letters*, 3(2), 222–226.

Meng, X., Currit, N., and Zhao, K. (2010). "Ground filtering algorithms for airborne LiDAR data: A review of critical issues." *Remote Sensing*, 2(3), 833–860.

Meng, X., Wang, L., Silván-Cárdenas, J. L., and Currit, N. (2009). "A multidirectional ground filtering algorithm for airborne LIDAR." *ISPRS Journal of Photogrammetry and Remote Sensing*, 64(1), 117–124.

Mezian, M., Vallet, B., Soheilian, B., and Paparoditis, N. (2016). "Uncertainty Propagation for Terrestrial Mobile Laser Scanner." *International Archives of the Photogrammetry, Remote Sensing & Spatial Information Sciences*, 41, 331–335.

Miyazaki, H., Nagai, M., and Shibasaki, R. (2015). "Reviews of Geospatial Information Technology and Collaborative Data Delivery for Disaster Risk Management." *ISPRS International Journal of Geo-Information*, 4, 1936–1964.

Mongus, D., Lukač, N., and Žalik, B. (2014). "Ground and building extraction from lidar data based on differential morphological profiles and locally fitted surfaces." *ISPRS Journal of Photogrammetry and Remote Sensing*, 93, 145–156.

Morsdorf, F., Meier, E., Kötz, B., Itten, K. I., Dobbertin, M., and Allgöwer, B. (2004). "LIDAR-based geometric reconstruction of boreal type forest stands at single tree level for forest and wildland fire management." *Remote Sensing of Environment*, 92, 353–362.

Mountrakis, G., Im, J., and Ogole, C. (2011). "Support vector machines in remote sensing: A review." *ISPRS Journal of Photogrammetry and Remote Sensing*, 66(3), 247–259.

NSF (2015). "New U.S.-Japan collaborations bring Big Data approaches to disaster response." Available at: www.nsf.gov/news/news_summ.jsp?cntn_id=134609.

Olsen, M. J. (2013). Guidelines for the use of mobile LIDAR in transportation applications, *Transportation Research Board*, Vol. 748.

Olsen, M. J., Cheung, K. F., Yamazaki, Y., Butcher, S., Garlock, M., Yim, S., McGarity, S., Robertson, I., Burgos, L., and Young, Y. L. (2012). "Damage assessment of the 2010 Chile earthquake and tsunami using terrestrial laser scanning." *Earthquake Spectra*, 28, S179–S197.

Owens, J. D., Houston, M., Luebke, D., Green, S., Stone, J. E., and Phillips, J. C. (2008). "GPU computing." *Proceedings of the IEEE*, 96(5), 879–899.

Passalacqua, P., Do Trung, T., Foufoula-Georgiou, E., Sapiro, G., and Dietrich, W. E. (2010). "A geometric framework for channel network extraction from lidar: Nonlinear diffusion and geodesic paths." *Journal of Geophysical Research: Earth Surface*, 115(F1).

Peterson, K., Ziglar, J., and Rybski, P. E. (2008). "Fast feature detection and stochastic parameter estimation of road shape using multiple LIDAR." *Proceedings of Intelligent Robots and Systems, 2008. IROS 2008. IEEE/RSJ International Conference on, IEEE*, 612–619.

Plale, B., and Schwan, K. (2000). "dQCOB: Managing large data flows using dynamic embedded queries." *Proceedings of the Ninth International Symposium on High-Performance Distributed Computing*, IEEE, 263–270.

Plaza, A. J., and Chang, C.-I. (2007). *High Performance Computing in Remote Sensing*, Boca Raton and London, CRC Press, Boca Raton, FL.

Poulter, B., and Halpin, P. N. (2008). "Raster modelling of coastal flooding from sea-level rise." *International Journal of Geographical Information Science*, 22, 167–182.

Rabbani, T., Van Den Heuvel, F., and Vosselmann, G. (2006). "Segmentation of point clouds using smoothness constraint." *International Archives of Photogrammetry, Remote Sensing and Spatial Information Sciences*, 36(5), 248–253.

Rottensteiner, F., and Briese, C. (2002). "A new method for building extraction in urban areas from high-resolution LIDAR data." *International Archives of Photogrammetry Remote Sensing and Spatial Information Sciences*, 34, 295–301.

Rottensteiner, F., Trinder, J., Clode, S., and Kubik, K. (2005). *Automated Delineation of Roof Planes from Lidar Data*, In ISPRS Workshop Laser scanning 2005, 36, 221–226. ISPRS Workshop groups WG III/3, III/4, V/3.

Sabou, M., Bontcheva, K., and Scharl, A. (2012). "Crowdsourcing research opportunities: Lessons from natural language processing." *Proceedings of the 12th International Conference on Knowledge Management and Knowledge Technologies*, ACM, 17.

Sakaki, T., Okazaki, M., and Matsuo, Y. (2010). "Earthquake shakes Twitter users: Real-time event detection by social sensors." *ACM*, 851–860.

Sanders, B. F. (2007). "Evaluation of on-line DEMs for flood inundation modeling." *Advances in Water Resources*, 30, 1831–1843.

Schnabel, R., Wahl, R., and Klein, R. (2007). "Efficient Ransac for point-cloud shape detection." In *Computer Graphics Forum*, Wiley Online Library, 214–226.

Schön, B., Bertolotto, M., Laefer, D. F., and Morrish, S. (2009). ""Storage, manipulation, and visualization of LiDAR data." *Presented at the 3rd International Workshop: 3D Virtual Reconstruction and Visualization of Complex Architectures (3D-ARCH'2009)*, Trento, Italy, 25–28 February 2009: *International Archives of Photogrammetry, Remote Sensing and Spatial Information Sciences*: Volume XXXVIII-5/W1.

Shi, J. and Malik, J. (2000). "Normalized cuts and image segmentation." *IEEE Transactions on Pattern Analysis and Machine Intelligence*, 22(8), 888–905.

Shu, K., Sliva, A., Wang, S., Tang, J., and Liu, H. (2017). "Fake news detection on social media: A data mining perspective." *ACM SIGKDD Explorations Newsletter*, 19(1), 22–36.

Singh, A. (1989). "Review article digital change detection techniques using remotely-sensed data." *International Journal of Remote Sensing*, 10(6), 989–1003.

Son, H., Kim, C., and Kim, C. (2014). "Fully automated as-built 3D pipeline extraction method from laser-scanned data based on curvature computation." *Journal of Computing in Civil Engineering*, 29(4), B4014003.

Soudarissanane, S., Lindenbergh, R., Meneti, M., and Teunissen, P. (2011). "Scanning geometry: Influencing factor on the quality of terrestrial laser scanning points." *ISPRS Journal of Photogrammetry and Remote Sensing*, 66, 389–399.

Starcke, K., and Brand, M. (2012). "Decision making under stress: A selective review." *Neuroscience & Biobehavioral Reviews*, 36(4), 1228–1248.

Suveg, I., and Vosselman, G. (2004). "Reconstruction of 3D building models from aerial images and maps." *ISPRS Journal of Photogrammetry and Remote Sensing*, 58, 202–224.

Tan, W., Missier, P., Foster, I., Madduri, R., De Roure, D., and Goble, C. (2010). "A comparison of using Taverna and BPEL in building scientific workflows: The case of Cagrid." *Concurrency and Computation: Practice and Experience*, 22(9), 1098–1117.

Trinder, J., and Salah, M. (2012). "Aerial images and LiDAR data fusion for disaster change detection." *ISPRS Annals of the Photogrammetry, Remote Sensing and Spatial Information Sciences*, 1(4), 227–232.

Verma, V., Kumar, R., and Hsu, S. "3d building detection and modeling from aerial lidar data." *Proceedings of Computer Vision and Pattern Recognition*, 2006 *IEEE Computer Society Conference on, IEEE*, 2213–2220.

Voigt, S., Kemper, T., Riedlinger, T., Kiefl, R., Scholte, K., and Mehl, H. (2007). "Satellite image analysis for disaster and crisis-management support." *IEEE Transactions on Geoscience and Remote Sensing*, 45, 1520–1528.

Wähner, K. (2014). "Real-time stream processing as game changer in a big data world with hadoop and data warehouse." *InfoQ* (September 10, 2014).

Walker, N. D. (1996). "Satellite assessment of Mississippi River plume variability: Causes and predictability." *Remote Sensing of Environment*, 58, 21–35.

Yang, B., Dong, Z., Zhao, G., and Dai, W. (2015). "Hierarchical extraction of urban objects from mobile laser scanning data." *ISPRS Journal of Photogrammetry and Remote Sensing*, 99, 45–57.

Yang, C., Goodchild, M., Huang, Q., Nebert, D., Raskin, R., Xu, Y., Bambacus, M., and Fay, D. (2011). "Spatial cloud computing: How can the geospatial sciences use and help shape cloud computing?." *International Journal of Digital Earth*, 4(4), 305–329.

Yu, Y., Li, J., Guan, H., Wang, C., and Yu, J. (2015). "Semiautomated extraction of street light poles from mobile LiDAR point-clouds." *IEEE Transactions on Geoscience and Remote Sensing*, 53(3), 1374–1386.

Yuan, C. (2012). High performance computing for massive LiDAR data processing with optimized GPU parallel programming, The University of Texas at Dallas.

Zhang, K., Yan, J., and Chen, S.-C. (2006). "Automatic construction of building footprints from airborne LIDAR data." *IEEE Transactions on Geoscience and Remote Sensing*, 44, 2523–2533.

Zhou, Z., Gong, J., and Guo, M. (2015). "Image-based 3D reconstruction for post-hurricane residential building damage assessment." *Journal of Computing in Civil Engineering*, 30, 04015015.

Zhou, Z., Gong, J., Roda, A., and Farrag, K. (2016). "Multiresolution Change Analysis Framework for Postdisaster Assessment of Natural Gas Pipeline Risk." *Transportation Research Record: Journal of the Transportation Research Board* (2595), 29–39.

Zhu, Z., and Brilakis, I. (2007). "Comparison of civil infrastructure optical-based spatial data acquisition techniques." *Computing in Civil Engineering* 2007, 737–744.

4 Smart City Portrayal

Dynamic Visualization Applied to the Analysis of Underground Metro

Evgheni Polisciuc and Penousal Machado

CONTENTS

4.1 INTRODUCTION

Today, 54% of the world's population lives in urban areas and by 2050 this number is expected to reach 67% (Habitat 2016). However, these areas are currently faced with the challenges of growing car ownership, vehicle travel, and energy consumption. These issues can only be addressed by an integrated approach, involving energy, transportation, and information and communication technology.

Transportation systems are not only a key factor for economic sustainability and social welfare, but also a key dimension in the smart city agenda. A smart city is a place where the traditional networks and services are made more efficient with the use of digital and telecommunication technologies, for the benefit of its inhabitants and businesses. The wide deployment of pervasive computing devices (smartphones, GPS devices, digital cameras, etc.), increasing storage and processing capacity of computing, improvements in sensing and modeling capabilities, Internet of Things, near field communications or social media, just to name a few, offer opportunities for digital transformation in cities. There is an opportunity to use ubiquitous urban sensing, big data and analytics to better understand the real-time functioning of the cities, as well as to collect data necessary to estimate and calibrate the mobility and energy consumption models, providing powerful tools useful in several contexts like economic, social, and public policy. Smart transportation management systems should use technology and collect information about mobility patterns. Understanding personal travel patterns and modeling travel demand is essential to plan sustainable urban transportation systems to fulfill citizens' mobility needs. To do this effectively and timely, urban and transportation planners need a dynamic way to profile the movement of people and vehicles.

Profiling of urban movements has traditionally relied on the knowledge of land use patterns, but, while land use and transportation infrastructures tend to remain in the same form for a long time once they are put in place, urban movements, on the other, often change. Transport planning input information mostly comes from traditional survey methods that are expensive and time-consuming, giving planners only a picture of what has happened and needing an active involvement. In contrast, the wide deployment of pervasive computing devices and transport system records (e.g., ticket validation counts; traffic counts) provide unprecedented digital footprints, telling where and when people are, possibly in real-time, enabling dynamic mobility profiling. For instance, Trépanier et al. (2007) present a model to estimate the destination location for each individual boarding using the smart card data. More recently, Ma et al. developed an efficient data mining method to demonstrate the temporal travel patterns for public transport riders in Beijing (2013). With millions of users around the world, transportation researchers have also realized the potential of Social Network Analysis (SNA) for travel demand modeling and analysis (Carrasco et al. 2008). All these mobility data, together with modern techniques for geo-processing, SNA, data fusion and visualization offer new possibilities for deriving activity destinations and allowing us to link cyber and physical activities through user interactions, in ways that can be usefully incorporated into models of land use and transportation interactions. Mining individuals' mobility patterns is an emergent research area (Lin and Hsu 2014) for predicting future movements or destinations (Calabrese et al. 2013), traffic forecast (Krings et al. 2009) or city planning (Makse et al. 1995), for instance.

The field of information visualization provides methods and techniques that enable the exploration and analysis of georeferenced data. For instance, the *Graduated Symbol Map* – a common technique to visually depict point-based data, where the size encodes differences in the magnitude of the value being represented (e.g., counts of people), or the *Transit Map* – a technique to uniformly display complex node-link systems, such as metro network, or *Circular Graphs* – used to depict time-based data, just to name a few, are modern visualization techniques that can be efficiently used for visual analysis of spatial data (Slocum 2005). Most of the existing visualization methods are well-studied and proven approaches to specific data analysis tasks. However, in many situations – e.g., when multiple visualization techniques are applied simultaneously – conflicts may arise: visual clutter resulting from overlapping of elements; inconsistency of representation in different map layouts; etc. Moreover, the problem increases when individual techniques are applied in the form of static layers, creating additional inefficiency, which in dynamic environments can vary from simple point superposition to incompatibility of the map layouts (e.g., schematic versus exact map). Furthermore, the general visual depiction of spatio-temporal data is still a challenging problem, which has been addressed by many researchers of the field (Andrienko and Andrienko 2006).

With that said, the present work focuses on multiple aspects of integrating different visual techniques within a dynamic environment, (i) making them to complement each other, (ii) providing different perspectives for reading the same data (mainly through the means of generalization), and (iii) reducing redundant information and the number of graphical elements. For running the experiments, we implemented an interactive application, which has been developed taking into account general perceptual and cognitive properties of the graphic system discussed later in this chapter. The interactive application runs in real-time, providing means to observe changes in daily usage of the metro network from different perspectives (e.g., detection of overloaded areas of the network, analysis of the usage of individual metro stations, observation of the daily patterns of entire network, etc.). We showcase the application with the quantitative spatio-temporal data related to the metro stations of London – statistical information regarding the counts of passengers (e.g., entries and exits per station), aggregated by 15 minutes intervals. Ultimately, the proposed approach is based on the principles described by Munzener – identify and analyze the real-world problem; design a visualization system, validate the results, and reflect upon lessons learned (2014).

In the following section, we overview background work. Then, we detail our model for point representation, starting with the concept, then color-coding, and then proceed with implementation. Next, the map transformations, such as map morphing and linking line displacement, are detailed. This is followed by a description of the graphical user interface. Finally, we provide a critical discussion about the visualization and draw some conclusions.

4.2 BACKGROUND AND RELATED WORK

Taking into account the nature of the data at hand, and the modern techniques to visualize it, we considered the following topics related to our work – graduated symbol and transit maps, circular time series visualization, dynamic visualization, and geographic generalization.

4.2.1 Point Representation

Much research was done in multivariate data visualization, focusing primarily on visual clutter reduction through the means of diverse techniques including filtering (Shneiderman 1996) and clustering (Adrienko and Adrienko 2011). In the context of urban mobility clustering techniques vary from simple K-means (Ferreira et al. 2013) to sophisticated interactive *Kohonen Maps* (aka SOM) (Schreck et al. 2009. Ellis and Dix (2007) have introduced a taxonomy, with multiple examples, of the methods for visual clutter reduction, which among many includes point/line displacement, topological distortion, dimensional reordering, space-filling and pixel-plotting for spatial representation, animation for temporal depiction, or sampling, change in point size or opacity with the addition of the already mentioned techniques for the appearance of the data items. In this work, we rather focus on techniques for improving the emphasis and understanding of relevant information, mainly, through the means of exaggeration of point representation methods.

Point representation is a visualization technique to represent quantitative data on maps, typically known as *Graduated Symbol Map* (Slocum 2005). According to Jacques Bertin (2010) a point represents a location on the map and has no theoretical area, whereas the graphical elements that render a point can vary in size, color, etc. In point representation, the graphical elements can overlap. Consequently, when a large number of points is considered the visualization may become illegible due to visual clutter. One of the possible solution, suggested by Bertin (2010), is the superposition of the small elements over the larger ones. For example, in Young and Bilton (2010) the authors used *Graduated Symbol Map* to examine the New York Times' web site, and mobile traffic data, nationwide and around the world. In the produced visualization, each circle represents a number of accesses to the sites, aggregated by one-minute interval. The yellow circles represent traffic to the website, while the red circles represent traffic to the mobile site. Although, the researchers discovered some interesting patterns, overall the visualization presents an elevated degree of visual clutter. Clutter reduction in point representation is an open problem, and innumerous techniques that tackles this problem have been developed. An example that is similar to the presented work combines generalization and interaction to depict data with varying degrees of point overlapping (Polisciuc et al. 2013).

4.2.2 Geographic Generalization

In geographic visualization, one of the used techniques to represent high-dense data is data abstraction, which is in some sense similar to geographic generalization (Bertin 2010). Due to limitations of human perception certain graphical elements are not distinguishable at higher levels of geographic zoom.* In this case, instead of direct representation, the concepts are generalized and then these are rendered with appropriate graphical elements. For instance, the work of Kim et al. (2014) focuses on a multivariate abstraction for representing spatio-temporal data. The data is estimated from a set of spatio-temporal events and visualized by means of bristled lines of the underlying vector map.

In some cases, abstraction is achieved through distortion of the representation. In a study done by Sips et al. (2004), abstraction results in the distortion of underlying map for the purpose of decreasing point overlapping. The higher the abstraction, the lower the level of overlapping. In contrast, distortion can be used as a mean to represent the data. For instance, in the SubMap project by Bujdosó et al. (2011) the vector maps of Budapest and Finland are distorted in response to high biking and Twitter activity, respectively. In the context of our work, we prefer the idea of exaggeration to distortion. As will be described later in this chapter, our approach does not distort the map, but exaggerates the representation highlighting the areas of attention, which are derived from the data.

4.2.3 Heatmap

Similar to the visualization technique known as heatmap (Slocum 2005) our approach is based on the idea of pseudo-elevation, this is, using color to represent a third dimension. In the case of heatmap, the elevation, which encodes the density of points as a rule, is represented by the gradient from green to red meaning low and high elevation, respectively. This technique is an efficient approach to visualize spatial data density, thanks to exaggeration that creates clear and distinguishable color ramps. The heatmap was successfully applied in the work of Fisher (2007). In his article the author presents Hotmap, which uses a heatmap to represent aggregate activity of the use of Microsoft's Live Search Map.

The use of elevation to represent spatial data for the purpose of urban mobility studies can be encountered in diverse applications. For instance, in the

* In the context of this work three levels of zoom are defined as follows: the general view – in this view all the considered geographic area is visible on the screen; the elementary views – in these views all the details are visible and distinguishable (e.g., streets, houses, etc.); and the intermediate views – as the name suggests, are the views that lay between the general and the elementary views.

project Real Time Rome (Calabrese and Ratti 2006) the cellphone activity was visualized by dividing the Rome urban area by a modular grid and by assigning a telecommunication traffic intensity to the elevation of each cell. The visual output is a 3D layer over the urban space that emphasizes the areas of attention similarly to heatmaps.

4.2.4 Circular Plot

There are different visual methods and techniques to analyze and characterize time-oriented data (see Aigner 2006; Chick et al 2003; Silva and Catarci 2000; Aigner et al. 2008). The main purpose of these techniques is to reveal periodic behavior, compare changes in continuous data in order to identify patterns, similarities and exceptions. Due to the nature of our dataset, one representative day of metro usage for weekdays and weekends, we focus on the methods to depict cyclic time. These methods can be divided in two main groups – representation in circle and representation using spirally shaped axis. The first method is usually applied when there is only one full cycle or when there are multiple attributes in the time series. In the second case the data attributes are mapped along the circumference and the time evolves from the center to the edge of the circle (Keim et al. 2004).

Visualizing time-oriented data on spiral is another popular technique. For instance, Spiral Graph (Weber et al. 2001) depicts cyclic features of data by using spirally shaped time axis, which clearly shows cyclical patterns if the cycle length is correctly aligned. Further this technique was improved by applying two-tone pseudo-coloring, which yields to a more accurate reading of the data (Tominksi and Schumann 2008). Also, time-oriented data can be visualized through the use of a swarm system applied on a spiral axis and with an expressive visual language to better convey information and visually engage the end user (Maçãs et al. 2015) Finally, interactive techniques can be used along with the spiral graphs, allowing the user to analyze a specific period of the time in more detail (Carlis and Konstan 1998; Tominski and Schumann 2008).

4.2.5 Schematic Map

Schematic map, more known as *Transit Schematic Map*, is a representation of geographic systems, such as roads and subway networks, using abstract graphic elements with the goal of clear communication of the topology of such systems. The main characteristics of schematic maps are the omission of all irrelevant details, the representation of relative position of geographical features and relations between them using abstract graphical elements (Guo 2011). A historical example is a Tube Map designed by Henry Beck in 1931 (Garland 1994). Beck depicted the London's underground network using its schematic representation. This map consists of the graphical elements that represent stations and straight-line segments connecting them. Lines are only vertical, horizontal or on 45-degree diagonal. Stations are differentiated between ordinary stations, marked with tick marks, and interchange stations, marked with diamonds sign.

More recent works describe techniques for automatic construction of schematic maps. For instance, the work of Cabello and Kreveld introduces an algorithm for automatic construction of schematic maps given the shape of the schematic connections and the distance between them (2002). The work by Bernhard analyzes the current London Underground Metro map and presents visual results for the geometric accuracy and distortion of maps, which can be used as a method for "geometric truth" of schematic maps (Jenny 2006).

4.3 POINT REPRESENTATION: DESIGN STUDY

In this work a simple visualization technique as point representation is revisited and expanded, taking into consideration current hardware, algorithms and data availability, and applied in the context of urban mobility. More precisely, we reconsidered a point representation of spatial data, where the quantities are encoded with the size of circles. Due to spatial data heterogeneity the point representation generates a large amount of overlapping points. Thus, the problem increases with the amount of data, resulting in high degrees of visual clutter, particularly in the general views of the map. To tackle this problem, we adapted a variation of the Metaball algorithm, and implemented it using *OpenGL Shading Language*. Finally, the model reacts to changes in zoom and transforms its representation accordingly, providing different degrees of graphical accuracy and modes of interpreting the visualization.

This section starts with the concept definition. First, we describe the underlying model, and then proceed with the formulation. The following subsection describes the color-coding for our model, with further formulation of the algorithm and implementation of the model based on hardware acceleration.

4.3.1 Concept and Formalization

In order to compare the proposed technique with a canonical point representation we started with the direct representation of data. The points were rendered by circular graphical elements with variable area. Each circle encodes the difference between entries and exists in the London Underground Metro at a given time interval. Positive and negative values are encoded with the red and green colors, respectively. As expected the dense areas contain numerous overlapping points that generate unwanted visual clutter. In the context of data exploration and analysis interaction techniques, such as zooming and panning, are used to approximate and navigate in the geographical space. These techniques allow the user to approximate highly dense areas and to analyze them in detail using closer views. However, the amount of visual clutter may be inappropriate for analytic and communication purposes.

As previously mentioned, a heatmap applied in geographical context enables the representation of high-dense data revealing the areas of interest. Moreover, this technique is efficient in conveying spatial information, although exaggerating its visual representation. The underlying idea for our approach was based on the visual exaggeration of the data in general view and the rigorous point representation in zoomed views. The combination of both features enables an efficient representation of high-dense data in any views, while keeping the same graphical elements and the same visual language throughout the views.

In order to combine the two graphical representations with the interactive aspects, the variable and constant parts of the representation were established. On the one hand, the general view must present strong detection and assembly characteristics, i.e., graphical elements should be displayed in a way that allows them to be easily detected and visually combined in meaningful groups. Exaggeration causes important elements to be distinct from other ones, revealing patterns and emphasizing important parts of the data. On the other hand, the zoomed views must have efficient visual estimation characteristics, i.e., the graphical elements must rigorously correspond to the data they encode. This allows the user to analyze the data in precise manner. Ultimately, fallowing the principles identified by Hall et al. (2016) the exaggeration, the zooming and the map transformations, which will be covered later, are used to

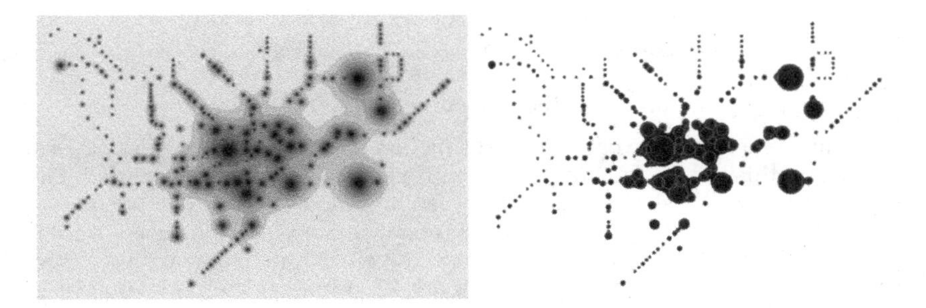

Figure 4.1 Image on the left – no threshold applied. Image on the right – threshold of 1 applied. The area of the red circles encodes corresponding data value. Images are generated with 303 given points, each one corresponding to a metro station. For the sake of demonstration of pseudo-elevation, the image on the left was fragmented using multi-band threshold, where white-to-black gradient represents the elevation from low to high.

create emphasis in visualization, in order to draw user's attention to the areas of interest.

Having that in mind, we employed an algorithm that generates visual artifacts with pseudo-elevation characteristics, similarly to heatmap. More precisely, a variation of Metaball, also known as blob algorithm, was used (Blinn 1982). Basically, this consists of a mountain-like surface where only the parts that are above a given threshold are rendered. In other words, Metaball defines a function, in our case in 2D space, which takes as input x and y coordinates of a point, and outputs a value. Then, the value becomes 1 if above a defined threshold or 0 if below. One of the common functions to define an underlying surface is a magnetic field equation $1/d^2$, where d is the distance between the given point and the point of the surface. One of the properties of this function is that the output never equals 0, meaning that every surface point is influenced by every given point. In other words, the points of the surface closer to input points have bigger elevation. The Figure 4.1 presents the visual outcome of Metaball algorithm with no threshold applied, image on the left, and with threshold of 1 applied, image on the right.

By comparing the Metaball output with the point representation (see Figure 4.1, image on the right) the following can be observed: the areas with high point density generate an elevated degree of positive exaggeration, while the areas with a small number of points generate negative exaggeration. By positive and negative exaggeration, we mean the following: the areas that encode a given value, sometimes referenced as a meaningful area, are bigger or smaller than expected; while positive exaggeration overestimates the data highlighting the areas with high deviations, negative exaggeration leads to underestimation of data, hiding information. In order to generate solely positive exaggeration, or no exaggeration at all, the points that are located inside the circles should have a value of pseudo-elevation greater than the defined threshold. In this case the total meaningful area created by Metaball algorithm would be equal or bigger than the one of the point representation. With that said, the elevation value V of each point of the surface p is computed as follows:

$$V_p = \sum_c \left(\frac{r_c}{\|c - p\|} \right)^b \tag{4.1}$$

where:

- r is the expected radius of the blob
- c is a given point
- b is a constant that controls the slope of the fall-off function

By controlling the slope of the fall-off function the amount of exaggeration can be changed, this is, how much the meaningful area is bigger than expected. This parameter also establishes a relation between precision/detail and expressiveness/abstraction of the representation. For instance, supposing the users want to visualize changes in the overall pattern without much detail, they would choose a lower value for b. The two images on Figure 4.2 exemplify the visual output resulting from the use of Equation 4.1 for different values of b. As will be established in Section 4.3.4, a value for $b = [2.3, 3.0]$ presents a good trade-of between exaggeration and accurate organic-looking visualization.

4.3.2 Color-Coding

This model comprises two types of colors – primary and secondary. The primary colors are red and green, and they encode the sign of the given values. Positive and negative values are represented by red and green colors, respectively. The secondary color is the yellow, and it emerges from the superposition of elements with primary colors (see Figure 4.3). The underlying idea for using secondary color comes from the problem of circle overlapping. That is, when two circles of the same size painted with different color are superposed, we rather paint with the secondary color, which clearly indicates that elements that represent different categorical values exist. The core idea is to make a direct blend of primary colors, in particular their red, green and blue components. A formal description of how this is accomplished is followed in the next subsection.

4.3.3 Implementation

Due to complexity of the algorithm, which is $O(n^m)$, where n is the number of given points and m is the number of points on the surface, we resorted to hardware acceleration, which enables the visualization to run in real-time. In order to reduce the number of communications between *RAM* and *GPU*, while rendering the data is pushed directly to the video card storing it in the *FloatBuffer*. This yields better performance, since the *GPU* to video card *RAM* speed is faster than the *GPU* to system *RAM*. The data array consists of the $[x_1, y_1, r_1, x_2, y_2, r_2, \ldots]$ sequence. The x and y are the Cartesian coordinates of each given point, and r is

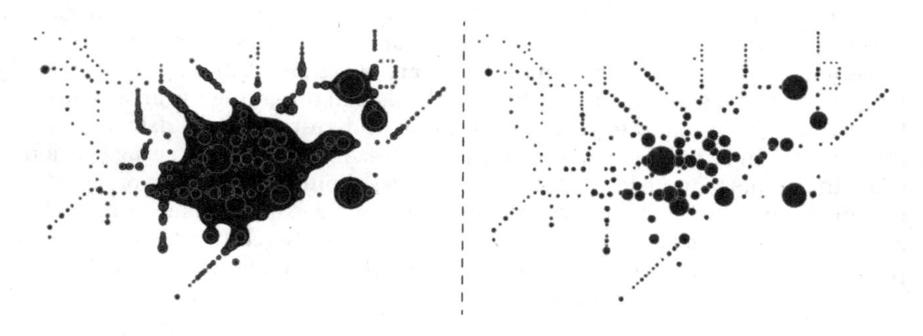

Figure 4.2 The effect of different slopes of the fall-off function – 1.0 and 5.0 for the images on the left and right, respectively.

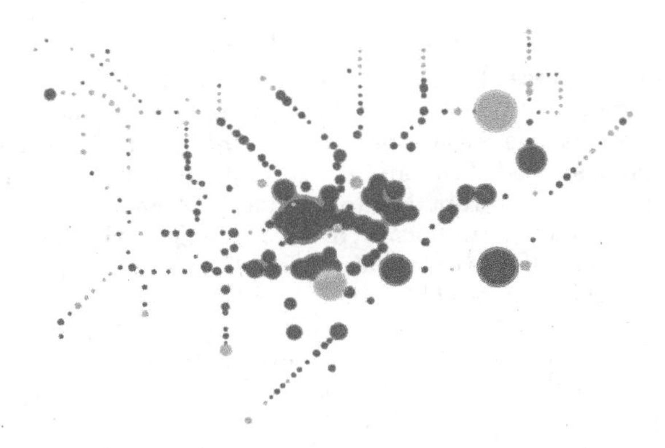

Figure 4.3 Color-coding for negative and positive values. Red and green depict positive and negative values, respectively, and yellow represents mix of both in overlapping locations.

the expected radius. The computation and rendering are both performed in the *GPU* using shaders.

The implementation of the fragment shader is presented in pseudo-code Algorithm 1. The function *influence()* is the implementation of Equation 4.1. The fragment shader runs this code for each pixel of the canvas and computes its color. Depending on the category of the given points the influence is accumulated to the corresponding color. For the sake of simplicity and since in our case there are only two categories, the sign of each class is bound directly with the color components – red and green for positive and negative values, respectively.

Algorithm 1 Implementation of the *main()* function in fragment shader

```
redComponent, greenComponent, alphaComponent ← 1.0
for all x, y, r in points do
    if r > 0 then
        redComponent ← redComponent + influence(x, y, abs(r))
    else
        greenComponent ← greenComponent + influence(x, y, abs(r))
    end if
end for
if redComponent < 1.0 and greenComponent < 1.0 then
    if max(redComponent, greenComponent) ≥ threshold then
        if redComponent > greenComponent then
            redComponent ← 1.0
            greenComponent ← 0.0
        else
            redComponent ← 0.0
            greenComponent ← 1.0
        end if
    else
        alphaComponent ← 0.0
    end if
end if

glFragColor ← [redComponent, redComponent, 0, alphaComponent]
```

4.3.4 Graphical Error

Graphical error is expressed as the difference between the area occupied by the rendered circles of the actual data and the area rendered by the Metaball algorithm. In other words, it is the additional pixels that create the trade-off between the exaggeration described above and the accurate representation. In order to understand how the fall-off function and dynamic zoom actually affects exaggeration, we measured the graphical error as a function of both components. The graphical error itself was computed as $E_p / T_p - C_p / T_p$, where E_p and C_p are the number of pixels painted by Metaball algorithm and the circles, respectively, and T_p is the total number of pixels in canvas. The plot of normalized graphical error as a function of zoom scale and fall-off value is shown in Figure 4.4. Additionally, we established a safe margin for the graphical error, which does not affect the perception of the data, neither the accuracy of the representation. Empirically, a safe margin in the range of [0.005, 0.02] corresponds to a fall-off value between 2.3 and 3.0. As can be observed from the figure, the higher graphical errors occur at lower levels of zoom, hence the established interval for the fall-off function. Ultimately, the behavior shown in the figure is what we

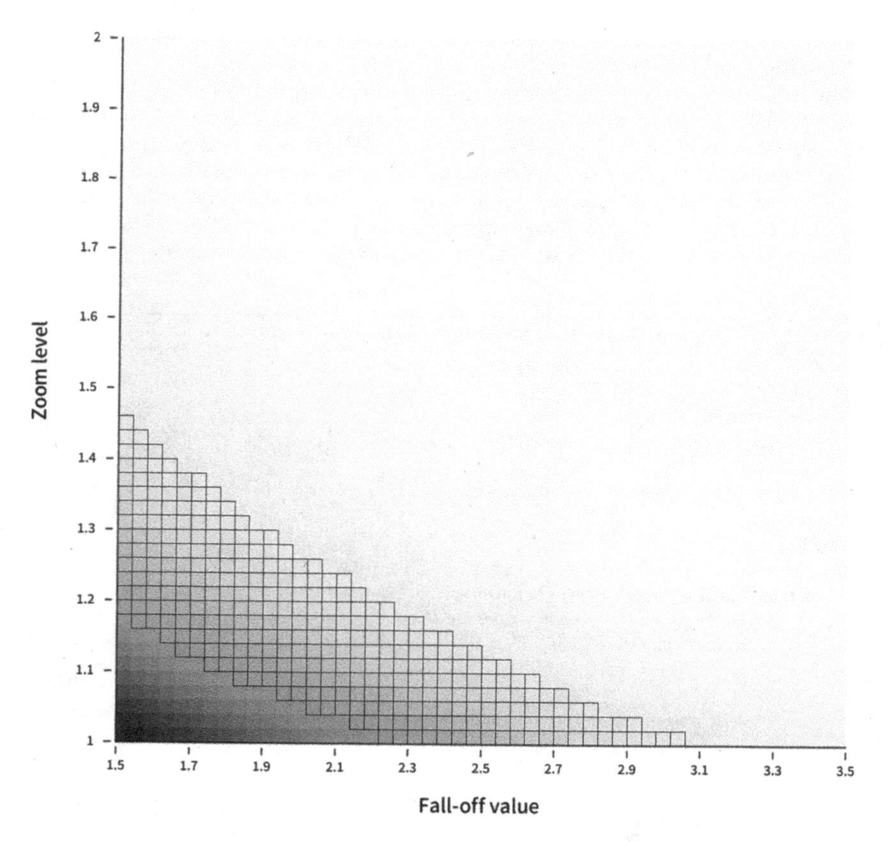

Figure 4.4 Graphical error as a function of zoom level and *b* value of the fall-off function. Blue color represents higher and yellow color lower error values. The highest and the lowest error values are 0.047 and –0.0057, respectively. The safe margins for graphical error are squared with black lines.

expected, since the model should present some degree of exaggeration at lower levels of zoom and transform into an accurate representation of data at higher zoom levels.

4.4 MAP TRANSFORMATIONS

The second core feature of this work is the transformations applied to the metro map in real-time and dynamic environment. These transformations are: transitions between two map formats – schematic and georeferenced metro maps; changes in line displacement in function of zoom scale; changes in data scaling. In the context of dynamic visualization animated transitions are a common practice nowadays (Heer and Robertson 2007). The changes in the visualization presented in an animated form facilitate their perception, therefore increasing the effectiveness of the representation (Robertson et al. 2008). The need for smooth transformation increases when multiple layouts of representation exist. In our case, the user should clearly perceive and correlate the location of stations in both map formats, in order to be able to track changes in data regardless of the zoom level. With that said, this section describes map morphing, edge displacement and the reactive data scale.

4.4.1 Map Morphing

In the case of the metro map, morphing consists of transiting from one map style (schematic map) to another (georeferenced map) and vice versa. Similar to conventional morphing applied in computer graphics, where each vertex of object A is mapped to the corresponding vertex of object B, we mapped stations and linking connections of both maps. In this section we begin by describing the map format and proceed with the description of the mapping techniques.

4.4.1.1 Metro Map Format

Due to their natures the schematic and real metro maps are stored in different formats. However, the common components, such as stops, lines, and connections, are stored in separate files. The schematic map is a direct transcription of the existing underground map. We developed an algorithm that parses the *SVG* image of the map, and stores it in a unique file with the following structure: *stops* – contain ID and coordinates in screen space; *curves* – describe curved connections of the map in the form of ID of the stops they connect and the coordinates of the vertices of the polyline; *lines* – contain ID and a set of coordinates that describe each metro line in screen space. Finally, the files *stations* and *connections* are structured according to a traditional directional graph representation – stations contain information regarding the ID, coordinates, in our case in geographic space, and additional information, while connections contain information about origin, target stop and the metro line number. Renderings of both maps are presented in Figure 4.5.

4.4.1.2 Station to Station Transition

In order to make the seamless transition between two maps perceptible, we applied a linear interpolation strategy. Since both maps lay in different coordinate spaces, the very first step is to align bounding boxes, origin axis and the scale. Both bounding boxes were aligned empirically in the center of the town, as this is the densest area. Having bounding boxes aligned, the transition between stations of both maps is performed using a linear interpolation of each of the components of the coordinates. The parameter, alpha value, that controls the transition lies in the range [10, 1], and is bound to a zoom range. More precisely, the transition starts slightly after the first zoom levels, and ends after a

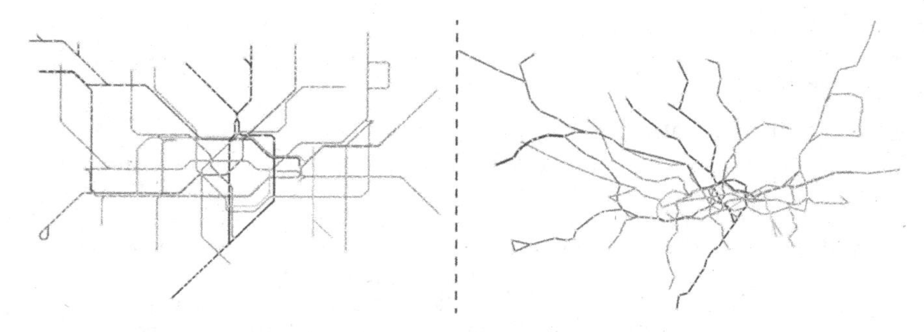

Figure 4.5 Representation of schematic (left) and georeferenced (right) maps. Colors correspond to the ones that are established in actual schematic map of London metro.

few more ticks of zoom. The presentation of actual values is unreasonable, since we used an arbitrary relative scale for the zoom levels.

4.4.1.3 Connection to Connection Transition

As can be seen in Figure 4.5 (image on the left), the schematic map comprises lines that are drawn in the form of a curve. Since the lines of the schematic map are drawn at 0°, 90°, and 45° angles and the stations are equidistant, it is inevitable to end-up in such drawing style. Thus, the transition between the straight lines of the real metro map and the curved lines of the schematic map becomes challenging. In order to solve this problem, we proceed with the following strategy: for each connection that has a corresponding curved line, first, subdivide the line with the segments proportionally equal to the segments of the curved line, and then similarly to the station transition, make transition between each individual vertex of a line. Connection to connection transition uses the same alpha value, so the entire map is transformed at the same time and scale.

4.4.2 Splitting Overlapped Segments

The georeferenced network presents a particular issue, which comes from the nature of data. This issue is the overlapping of segments that connect pairs of stations that share multiple metro lines. Moreover, the dataset does not describe the exact location of connecting lines, which is dictated solely by the geographical locations of stations, causing them to overlap when representing graphically.

To overcome this issue the following approach was applied: to compute corresponding render locations for each station we look for adjacent segments, identified by the stations they connect, distribute them evenly along the perpendicular of a common axis and center them on that axis, which connects both stations location; finally, the order at which the segments are distributed is dictated by the number of the metro line they represent. The order direction is not important, although the same ordering within adjacent segments should be retained to preserve continuous representation of the segments that belong to the same line. Ultimately, the displacement offset corresponds to a fixed thickness of rendered line (*fixedWidth*), such that each segment is rendered right next to each other without any gaps.

That is, given a set of stations $S = \{s_1, s_2, \ldots, s_n\}$ the algorithm proceeds as shown by pseudo-code in Algorithm 2. The function *getAdjacentConnections()*

proceeds with sorting the array of connections, comparing the pairs of IDs of stations they connect, and returns an array of arrays, each one containing segments that connect same pair of stations. The *offsetVector* is used to compute the displacement of each segment. Finally, the *perpendicular* vector, with a magnitude equal to half of the sum of the widths of the segments within each group, defines starting locations from which each segment is displaced.

Algorithm 2 Pseudo-code for the segment displacement algorithm

```
for all s in S do
    if s.connections > 1 then
        {sort connections by station IDs and create groups of adjacent connections}
        groupsAdjacentConnections ← getAdjacentConnections(s.connections)
        for all adjacentConnections in groupsAdjacentConnections do
            {sort connections within each group by line ID}
            sortByLineID(adjacentConnections)

            perpendicular ← adjacentConnections[0].perpendicular
            perpendicular.magnitude(−(fixedWidth * (adjacentConnections.length − 1))/2)

            offsetVector ← perpendicular
            offsetVector.magnitude(fixedWidth)

            renderLocationA ← perpendicular + adjacendConnections[0].stationA.location
            renderLocationB ← perpendicular + adjacendConnections[0].stationB.location
            for all connection in adjacentConnections do
                connection.renderLocationA ← renderLocationA
                connection.renderLocationB ← renderLocationB

                renderLocationA ← renderLocationA + offsetVector
                renderLocationB ← renderLocationB + offsetVector
            end for
        end for
    end if
end for
```

4.4.3 Reactive Circle Scaling

In this project we opt for relative graphical scaling, i.e., the graphical components, in our case the size of the circles, are scaled in proportion to the data being displayed in the canvas at a given time interval. This means that only the displayed objects can be correctly compared with each other. This approach solves the problem of representing values with big difference in magnitude, i.e., when the representation of high values hinders the perception of the smaller ones, due to the effect of scale.

We start by identifying the stations begin displayed at each interaction instance – zooming or panning. First, the bounding box, which is located at the origin of axis and equal to the canvas size, is projected onto the transformed map space. We take inverse of the transformation matrix and multiply it by the bounding box location and scale the size according to the matrix scale. Then, we check for the inclusion of each station within the bounding box, taking the maximum value in that region. Finally, the local maxima found is used to scale all the other values of the displayed stations. Figure 4.6 exemplifies the reactive circle scaling.

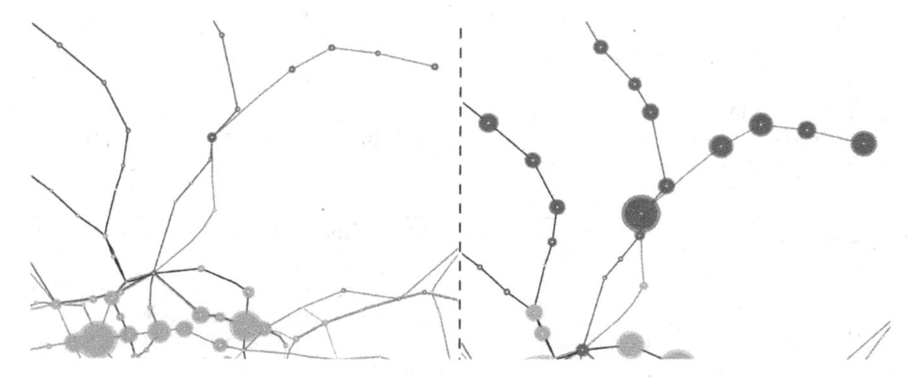

Figure 4.6 Comparison between different view instances and the corresponding response from visualization. Image on the right represents the map pushed down along *Y* axis compared to the image on the left. All the relative proportions are retained, except the local maxima.

4.5 INTERACTIVE APPLICATION

The previously described point representation and the map transformations are the core of our visualization method. To help users interactively explore and analyze movement of inhabitants through the metro network we designed and developed interactive graphical user interface. As shown in the Figure 4.7, the interface consists of multiple components, which are described in this section.

4.5.1 General Overview

Exploring and analyzing the use of a complex metro network often consist of multiple tasks, such as: per station daily usage analysis; comparison of the users' flow among metro stations; detection of overused regions or stations; etc. The interactive application depicted in the Figure 4.7 provides means to fulfill these tasks. At the upper left corner, the user can find a circular diagram. This provides means to navigate in time and depicts data regarding a daily usage of the selected metro station. The main window depicts the core visualization. The user can zoom and pan the map, as well as click on the stations to get additional information. Zooming in on the map behaves as described earlier – the schematic map transforms into an accurate georeferenced representation. At the end of the transformation a base map is added, allowing the user to better orient in space. The interactive application can be seen in action in the following link https://cdv.dei.uc.pt/2017/metro-analyzer.mov.

4.5.2 Circular Diagram

The circular diagram, found in the upper left corner of the window (see Figure 4.7), serves two main functions – to allow the user to navigate in time and show data for the selected station. The diagram is presented in the circular form and is subdivided by 24 equal parts, each one corresponding to one hour of the day. By clicking on the diagram, the application jumps to the corresponding time interval, updating the visualization accordingly. The pointer is positioned according to the selected time interval and the label is updated as well. Finally, when user selects a station on the map its name appears centered on the top of the diagram, and its data is depicted by a circular plot.

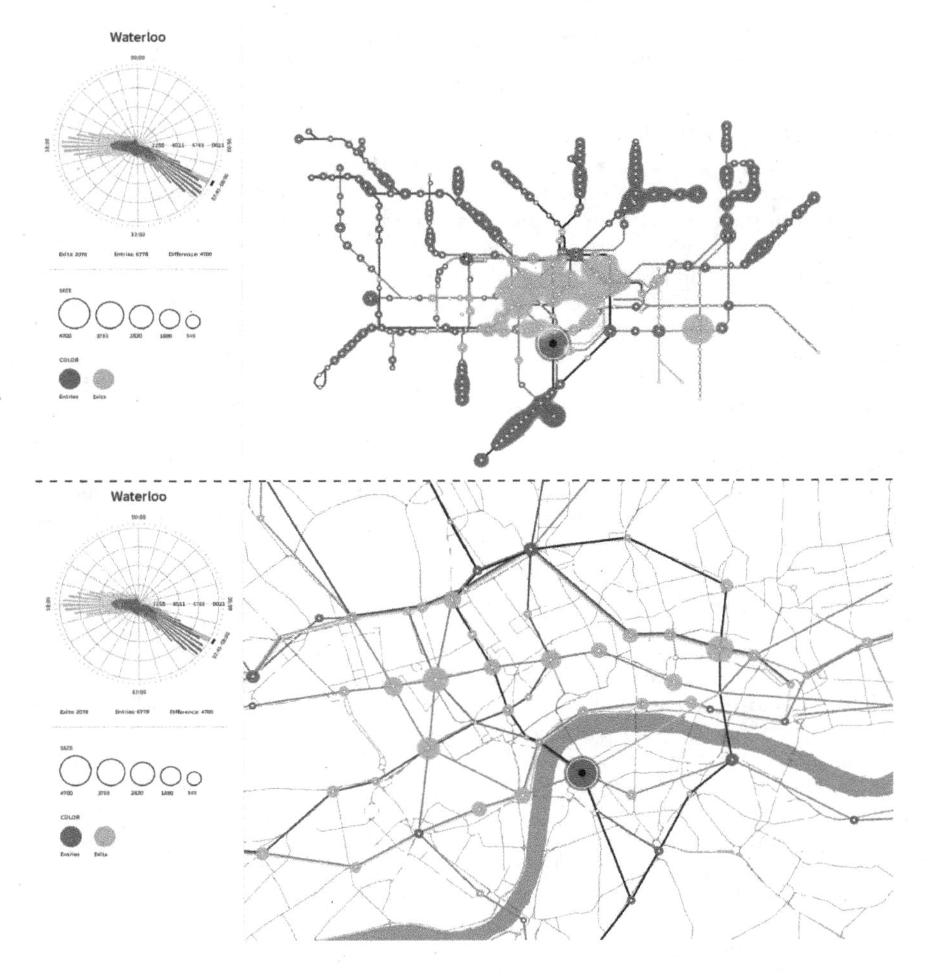

Figure 4.7 Interactive application. Image on the top shows fully zoomed out map, and image in the bottom shows end of the transformation of the map. In both representations, the same station and time interval are selected.

Regarding the representation of data in the circular plot, we considered both *entries* and *exits* per metro station. Similar to reactive data scaling described earlier, the circular plot shows relative data, i.e., the representation is only comparable within the selected station and time interval. This is done because we want to separate different instances of the analysis, as well as diminish redundant information, making each part of the visualization complement each other. Ultimately, the circular plot should provide a broad picture of the daily usage of the station, rather than allowing the comparison of the values among stations. Nevertheless, the ranges for the values are shown in the visualization.

Entries and exits of a metro station are depicted using bars. The height of each one represents the counts of metro users at the selected station, and the red and green colors represent entries and exits, respectively. The grid with four divisions provides visual cues that help analyzing the data. The labels are updated according to the maximum value found throughout the day at selected station. Finally, for each time interval the color of the predominant value corresponds to the color on the map.

4.6 CRITICAL DISCUSSION

This section provides a qualitative evaluation based on Tufte's (2001) principles on graphical excellence and integrity, as well as a theoretical evaluation taking into account the properties of graphic system proposed by Bertin (2010). In general, the presented visualization follows the main principle of graphical excellence, which gives to the user elevated number of ideas in shortest time with the least "ink" in smallest space. First, in regarding the circular diagram, this is based on the traditional method of visualizing time-oriented data. As the color hue variable, also used in the main visualization, claims to be the most efficient for representing categorical data, this is appropriate usage as the user want to make a clear distinction between entries and exists. The use of a circular grid and the labels for temporal and data divisions encourage graphical clarity and the precision of reading the plot.

The main map provides multiple readings, such as pattern detection in temporal and spatial domains, low-level station analysis, and comparison of entries and exits, and corresponding difference, among the stations. The usage of the size variable is also appropriate, since according to Bertin this visual element and the plane are the only variables that can be used quantitatively. However, the aggregation technique introduces graphical error in the rendering of blobs, that can lead to incorrect estimation of the values, if not used as the model was designed – observe and identify areas of interest in the general views and analyze in closed views. Also, the rendering of the blobs results in additional "ink," which may conflict with the principle of data-ink minimization – present the largest amount of data with least amount of "ink." On the other hand, the aggregation technique reduces overall visual clutter, which perfectly fits into Tufte's principle on data density. Last but not least, in the close views, a reference base map is added, such that the user can accurately and rapidly localize each station in the city. Finally, the visualization as a whole reveals multiple details and provides multiple perspectives over the data, which ensures Tufte's graphical excellence. Additionally, the visualization is accompanied with the dynamically updated legend that assists with the reading of the graphic and decoding data.

4.7 CONCLUSIONS

In this article we presented a visualization application that combines characteristics such as pattern detection and low-level analysis of urban mobility and integrates multiple visualization techniques in a dynamic environment. While zooming, the model adjusts, such that the representation is efficient at any zoom scale. Our approach tackles the overlap problem by occluding unnecessary visual information in general views, while mixing the primary colors, which allows detection of occluded elements of different categories. The extended Metaball algorithm, implemented with recursion to hardware acceleration, is an efficient and effective solution to represent high-dense point-based geographic data in real-time. We presented the concept behind our model and the algorithm that computes the blobs from a set of points on geographic space. We improved aesthetic aspects of point representation by the means of exaggeration, which results in an organic-looking visual outcome. By changing the slope parameter of the fall-off function, the model can be adjusted to find a good trade-off between expressive visual artifacts and precise representation. Finally, we demonstrated how graphical error behaves in relation to zoom scale and fall-off function.

The second core feature of this work is the transformations applied to the representation of the metro map, which are essential for the effectiveness of the perception of changes in visualization. In our case, the user should clearly perceive and correlate the location of stations in both map formats while zooming. We presented methods for station to station and connection to connection transitions, link displacement, and reactive data scaling. These provide the user with the different perspectives for reading the same data, namely: detection of overloaded areas of the network at lower zoom scales in a visual clutter free display; and detailed analysis of station usage at higher zoom levels though an accurate representation. Finally, reactive data scaling partially solves the problem of representing values with big difference in magnitude, i.e., when the representation of high values hinders the perception of the smaller ones, due to the effect scale.

The interactive application presented provides the necessary tools to explore geographic time series. The circular diagram allows the users to navigate in time, giving them a first general temporal overview of the data, and then allowing them to move directly to the interesting time interval. Also, the patterns of metro network usage thought the day can be observed using the same graphical interface. Additionally, the circular diagram depicts the daily usage of the selected station. The map implements zooming and panning, translating these actions into a meaningful transformation of the visualization. Finally, each station is selectable, providing the user with additional information on demand.

ACKNOWLEDGMENTS

We would like to thank Rui Gomes and Catarina Maçãs for their contributions to this work. Also, we would like to thank the reviewers for their useful comments. This project has been supported by Fundação para a Ciência e Tecnologia (FCT), Portugal, under the grant SFRH/BD/109745/2015.

REFERENCES

Adrienko, N. and Adrienko, G.: Spatial generalization and aggregation of massive movement data. *IEEE Transactions on Visualization and Computer Graphics* 17(2), 205–219 (2011).

Aigner, W.: Visualization of time and time-oriented information: challenges and conceptual design. Ph.D. Thesis, Vienna University of Technology, Institute of Software Technology and Interactive Systems (2006).

Aigner, W., Miksch, S., Müller, W., Schumann, H., and Tominski, C.: Visual methods for analyzing time-oriented data. *IEEE Transactions on Visualization and Computer Graphics* 14(1), 47–60 (2008).

Andrienko, N., Andrienko, G.: *Exploratory Analysis of Spatial and Temporal Data: A Systematic Approach*. Springer Science & Business Media, Berlin, Germany, (2006).

Bertin, J.: Semiology of graphics: diagrams, networks, maps. ESRI Press. Distributed by In-gram Publisher Services, Redlands, CA (2010).

Blinn, J.F.: A generalization of algebraic surface drawing. *ACM Transactions on Graphics (TOG)* 1(3), 235–256 (1982).

Bujdosó, A., Feles, D., Gergely, K., Kiss, L., Füredi, G., Megyer, L., Véhmann, F.: Submap: visualizing locative and time-based data on distorted maps. (2011) http://submap.kibu.hu/.

Cabello, S. and van Kreveld, M.: Schematic networks: an algorithm and its implementation. In: *Advances in Spatial Data Handling*, In Advances in Spatial Data Handling. Springer, Berlin, Heidelberg, 475–486 (2002).

Calabrese, F., Diao, M., DiLorenzo, G., Ferreira, J., and Ratti, C.: Understanding individual mobility patterns from urban sensing data: a mobile phone trace example. *Transportation Research Part C: Emerging Technologies* 26, 301–313 (2013).

Calabrese, F., Ratti, C.: Real time rome. *Networks and Communication Studies* 20(3–4), 247–258 (2006).

Carlis, J.V. and Konstan, J.A.: Interactive visualization of serial periodic data. In: Proceedings of the 11th annual ACM symposium On user interface software and technology, 29–38. *ACM*, San Francisco, CA, (1998).

Carrasco, J.A., Hogan, B., Wellman, B., and Miller, E.J.: Collecting social network data to study social activity-travel behavior: an egocentric approach. *Environment and Planning B: Planning and Design* 35(6), 961–980 (2008).

Chick, S., Sánchez, P., Ferrin, D., and Morrice, D.: Visualization methods for time-dependent data – an overview. In: Proceedings of the 2003 winter simulation conference (2003).

Ellis, G., Dix, A.: A taxonomy of clutter reduction for information visualisation. *IEEE Transactions on Visualization and Computer Graphics* 13(6), 1216–1223, New Orleans, LA (2007).

Ferreira, N., Klosowski, J.T., Scheidegger, C.E., and Silva, C.T.: Vector field k-means: clustering trajectories by fitting multiple vector fields. In: *Computer Graphics Forum*, 32, 201–210. Wiley Online Library (2013).

Fisher, D.: Hotmap: Looking at geographic attention. *IEEE Transactions on Visualization and Computer Graphics* 13(6), 1184–1191, Blackwell Publishing Ltd, Oxford, UK (2007).

Garland, K.: Mr Beck's underground map. Capital Transport, Middlesex (1994).

Guo, Z.: Mind the map! the impact of transit maps on path choice in public transit. *Transportation Research Part A: Policy and Practice* 45(7), 625–639 (2011).

Habitat, U.: Urbanization and development emerging futures. World cities report. Quito, Ecuador, (2016).

Hall, K.W., Perin, C., Kusalik, P.G., Gutwin, C., and Carpendale, S.: Formalizing emphasis in information visualization. In: *Computer Graphics Forum*, Vol. 35, pp. 717–737. Wiley Online Library, The Eurographs Association & John Wiley & Sons, Ltd. (2016).

Heer, J., Robertson, G.: Animated transitions in statistical data graphics. *IEEE Transactions on Visualization and Computer Graphics* 13(6), 1240–1247 (2007).

Jenny, B.: Geometric distortion of schematic network maps. *Bulletin of the Society of Cartographers* 40(1), 15–18 (2006).

Keim, D.A., Schneidewind, J., and Sips, M.: Circleview: A new approach for visualizing time-related multidimensional data sets. In: Proceedings of the working conference on advanced visual interfaces, pp. 179–182. *ACM*, Gallipoli, Italy (2004).

Kim, S., Maciejewski, R., Malik, A., Jang, Y., and Ebert, D.S., Isenberg, T.: Bristlemaps: a multivariate abstraction technique for geovisualization. *IEEE Transactions on Visualization and Computer Graphics* 19(9), 1438–1454 (2013).

Krings, G., Calabrese, F., Ratti, C., and Blondel, V.D.: Urban gravity: a model for inter-city telecommunication flows. *Journal of Statistical Mechanics: Theory and Experiment* 2009(07), L07,003 (2009).

Lin, M. and Hsu, W.J.: Mining gps data for mobility patterns: A survey. *Pervasive and Mobile Computing* 12, 1–16 (2014).

Ma, X., Wu, Y.J., Wang, Y., Chen, F., and Liu, J.: Mining smart card data for transit riders travel patterns. *Transportation Research Part C: Emerging Technologies* 36, 1–12 (2013).

Maçãs, C., Cruz, P., Martins, P., and Machado, P.: Swarm systems in the visualization of consumption patterns. In: Q. Yang, M. Wooldridge (eds.) Proceedings of the twenty-fourth international joint conference on artificial intelligence, *IJCAI* 2015, Buenos Aires, Argentina, July 25–31, 2015, pp. 2466–2472. AAAI Press (2015).

Makse, H.A., Havlin, S., and Stanley, H.E.: Modelling urban growth patterns. *Nature* 377(6550), 608–612 (1995).

Munzner, T.: *Visualization Analysis and Design*. CRC Press (2014).

Polisciuc, E., Alves, A., Bento, C., and Machado, P.: Visualizing urban mobility. In: ACM SIGGRAPH 2013 posters, p. 115. ACM (2013).

Robertson, G., Fernandez, R., Fisher, D., Lee, B., and Stasko, J.: Effectiveness of animation in trend visualization. *IEEE Transactions on Visualization and Computer Graphics* 14(6), Anaheim, CA (2008).

Schreck, T., Bernard, J., Von Landesberger, T., and Kohlhammer, J.: Visual cluster analysis of trajectory data with interactive kohonen maps. *Information Visualization* 8(1), 14–29 (2009).

Shneiderman, B.: The eyes have it: a task by data type taxonomy for information visualiztions. In: Proceedings of the IEEE symposium on visual languages, 1996, pp. 336–343. *IEEE*, Boulder, CO (1996).

Silva, S.F. and Catarci, T.: Visualization of linear time-oriented data: a survey. In: Proceedings of the first international conference on web information systems engineering, 2000, Washington, DC, Vol. 1, pp. 310–319. *WIEEE* (2000).

Sips, M., Keim, D.A., Panse, C., and Schneidewind, J.: Geo-spatial data viewer: from familiar land-covering to arbitrary distorted geo-spatial quadtree maps. *UNION Agency*, Plzen, Czech Republic (2004).

Slocum, T.: *Thematic Cartography and Geographic Visualization*. Pearson/Prentice Hall, Upper Saddle River, NJ (2005).

Tominski, C. and Schumann, H.: Enhanced interactive spiral display. In: SIGRAD 2008. The annual SIGRAD conference special theme: interaction, Stockholm; Sweden, November 27–28, 2008, pp. 53–56. Linköping University Electronic Press (2008).

Trépanier, M., Tranchant, N., and Chapleau, R.: Individual trip destination estimation in a transit smart card automated fare collection system. *Journal of Intelligent Transportation Systems* 11(1), 1–14 (2007).

Tufte, E.: The visual display of quantitative information. Graphics Press, Cheshire, CT (2001).

Weber, M., Alexa, M., and Müller, W.: Visualizing time-series on spirals. In: *IEEE Transactions on Visualization and Computer Graphics* 1, 7–14 (2001).

Young, M., Bilton, N.: A day in the life of the New York times. In: J. Steele, N. Iliinsky (eds.) *Beautiful Visualization: Looking At Data Through the Eyes of Experts*, chap. 16, pp. 271–290. O'Reilly Media, Inc. (2010).

5 Smart Bike-Sharing Systems for Smart Cities

Hesham A. Rakha, Mohammed Elhenawy, Huthaifa I. Ashqar, Mohammed H. Almannaa, and Ahmed Ghanem

CONTENTS

5.1 INTRODUCTION

In the next few decades, many traditional cities will be turned into smart cities, which are greener, safer, and faster. This transformation is supported by recent advances in information and communication technology (ICT) in addition to the expected fast spread of the Internet of Things and big data analytics. Smart cities will mitigate some of the negative impacts of traditional cities, which consume 75% of the world's resources and energy and produce 80% of the greenhouse gases (Mohanty et al. 2006). Smart cities have many components, including smart transportation. Smart transportation will integrate different transportation networks and allow them to work together so travelers and commuters can enjoy seamless multi-mode trips based on their preferences. Consequently, more commuters will be inspired to use public transportation systems and many traffic-related problems such as congestion will be relaxed.

The last mile problem is a pressing problem that needs to be solved in order for different transportation networks to work together efficiently. This problem is defined as "the short distance between home and public transit or transit stations and the workplace, which may be too far to walk." One solution to this problem is a bike-sharing system (BSS), which takes advantage of the availability of ICT and the BSS's data to smartly operate the network. Smart bike-sharing systems (SBSSs) use recent technologies to monitor the status of each station in the network, collect bike usage data and other relevant data, and use state-of-the-art algorithms to build predictive models, predict future bike availability, and find good solutions for the issue of imbalance in the distribution of bikes in order to guarantee users' satisfaction and meet their demand.

Due to relatively low capital and operational costs, as well as ease of installation, many cities in the United States are making investments in BSSs. A technical report distributed by the Bureau of Transportation in April 2016 indicated that there are 2,655 BSSs stations in 65 U.S. cities, and that 86.3% of these stations are connected to another means of scheduled public transportation (Contardo et al. 2012). These numbers show that the physical infrastructure for BSSs already exists and that they are good candidates for connecting different transportation networks. In 2013, San Francisco launched the Bay Area Bike Share System (now called the "Ford GoBike" BSS), a membership-based system providing 24-hours-per-day, 7-days-per-week self-service access to short-term rental bicycles. Members can check out a bicycle from a network of automated stations, ride to the station nearest their destination, and leave the bicycle safely locked for someone else to use (Bay Area Bike Share 2016). The Bay Area Bike Share is designed for short, quick trips, and as a result, additional fees apply for trips longer than 30 minutes. In this system, 70 bike stations connect users to transit, businesses, and other destinations in four areas: downtown San Francisco, Palo Alto, Mountain View, and downtown San Jose (Bay Area Bike Share 2016). Bay Area Bike Share is available to everyone 18 years and older with a credit or debit card. The system is designed to be used by commuters and tourists alike, whether they are trying to get across town at rush hour, traveling to and from Bay Area Rapid Transit and Caltrain stations, or pursuing daily activities (Bay Area Bike Share 2016).

However, these systems suffer from a central recurring problem: rebalancing. Rebalancing is a daily problem for operators, who have to find an efficient way to redistribute (i.e., rebalance) bikes from full stations to empty stations to meet expected demand patterns. This redistribution problem is a generalization of the well-known traveling salesman problem (TSP), which involves finding the shortest route passing through each of a collection of locations and then returning to a starting point. This problem was first proposed in (Schuijbroek et al. 2013) as a one-commodity pick-up and delivery traveling salesman problem (1-PDTSP). 1-PDTSP is NP-hard, so heuristic optimization techniques are applied to determine a near optimum tour (i.e., route).

Rebalancing can be classified as either static, dynamic, or incentivized. In both static and dynamic rebalancing, BSS operators usually use a fleet of trucks to perform the task. Static rebalancing is generally referred to in the literature as the static bicycle repositioning problem (SBRP). The common assumption of SBRP algorithms is that the number of bikes at each station either remains the same or changes slightly, and does not affect the rebalancing outcome. Thus, demand prediction is needed to check the validity of this assumption. The dynamic bicycle repositioning problem (DBRP) assumes that moving bikes will have a significant impact on BSS user demand, which will affect the rebalancing outcome. As such, demand predictions have to be input into the algorithm for solving the DBRP so that they are incorporated into the solution. Incentivized rebalancing is based on providing BSS users with incentives to contribute to the system rebalancing. The BSS sends control signals to users suggesting slight changes to their planned journeys, providing them with alternate routes, or offering the option to return bikes for system credit. These suggestions will depend on the demand prediction at stations near the destination station of the planned trip.

In this chapter, we make three contributions to the literature. First, we adopt state-of-the-art machine learning techniques to build predictive models of the bike availability at each station in the BSS. The built models are compared in terms of mean absolute errors, and we identify which algorithm is suitable for which condition. Second, we propose a new supervised clustering algorithm to provide a global view of network-wide bike availability across stations. To do so, we developed a novel supervised clustering algorithm that is built using two well-known algorithms, namely the Gale-Shapley student-optimal college admission (CA) algorithm (Gale and Shapley 1962) and the K-median algorithm. The obtained global view of the BSS shows that some of the low-demand stations are located next to high-demand stations, making it possible to balance the system at minimal cost and effort. Third, we propose a new static rebalancing algorithm based on the deferred acceptance algorithm, a well-known game theory algorithm consisting of two phases: tour construction and tour improvement. The proposed algorithm models the rebalancing problem as two disjoint sets of players. Each player in these sets has their own objective function, which is used to build the preference list for players on the other side.

In terms of the chapter layout, following the introduction, we will briefly discuss the related work to the three main contributions of this chapter. A description of the data used in this chapter follows the related work. Then, each of our contributions and its related results will be discussed in a separate section. Finally the summary findings and conclusions on the work are presented.

5.2 RELATED WORK

In this section, we discuss recent research on bike availability prediction, clustering bike availability data, and rebalancing problems. We will discuss the prediction-related work first, then the usage of cluster analysis, and finally the previous research on rebalancing.

5.2.1 Bike Availability Prediction

Modeling bike sharing data is an area of significant research interest. Proposed models have relied on various features, including time, weather, the built environment, and transportation infrastructure. In general, the main goals of these models have been to boost the redistribution operation (Contardo et al. 2012; Schujibroek et al. 2013; Raviv et al. 2013), to gain new insights into and correlations between bike demand and other factors (Daddio 2012; Wang et al. 2015; Rudloff and Lackner 2013; Rixey 2013), and to support policy makers and managers in making optimized decisions (Daddio 2012; Vogel et al. 2011). In general, the prediction models can be classified into two broad classes, namely statistical models and machine learning models.

Many studies have been performed to predict the availability of bikes by using time series analysis. Rixey used a multivariate linear regression analysis to study station-level BSS ridership, investigating the correlation between BSS ridership and the following factors: population density; retail job density; bike, walk, and transit commuters; median income; education; presence of bikeways; nonwhite population (negative association); days of precipitation (negative association); and proximity to a network of other BSS stations (2013). The author found that demographics, the built environment, and access to a comprehensive network of stations were critical factors in supporting ridership.

Froehlich, Neumann, and Oliver used four predictive models to predict the number of available bikes at each station: last value, historical mean, historical trend, and Bayesian networks (2009). Results showed that historical mean predictor performed worst of all, while Bayesian networks outperformed all the other models. They also highlighted the importance of giving more weight to the most recent observations. Moreover, two methods for time series analysis, autoregressive-moving average (ARMA), and autoregressive integrated moving average (ARIMA), have also been used to predict the number of available bikes/docks at each bike station. Kaltenbrunner et al. adopted ARMA (2010); Yoon et al. proposed a modified ARIMA model considering spatial interaction and temporal factors (2012). However, Gallop et al. used continuous and year-round hourly bicycle counts and weather data to model bicycle traffic in Vancouver, Canada (2011). That study used a seasonal ARIMA analysis to account for the complex serial correlation patterns in the error terms and tested the model against actual bicycle traffic counts. The results demonstrated that the weather had a significant and important impact on bike usage. The authors found that the weather data (i.e., temperature, rain, humidity, and clearness) were generally significant and that temperature and rain, specifically, had an important effect.

On the other hand, few studies have used machine learning to model bike sharing data. One of the characteristics of transportation-related datasets is that they are often very large. It is therefore advantageous to implement machine learning to identify potential explanatory variables (Vogel et al. 2011). Moreover, when a model contains a large number of predictors, it becomes more complex and overfitting can occur. To address this, different algorithms have been used to predict bike availability in a BSS, such as random forest (RF), support vector machine, and gradient-boosted tree (GBT) (Yin et al. 2012; Du et al. 2014; Dias et al. 2015; Hernández-Pérez and Salazar-González 2004a; Giot and Cherrier 2014; Ashqar et al. 2017). The authors of the four studies in (Yin et al. 2012; Du et al. 2014; Hernández-Pérez and Salazar-González 2004a; Giot and Cherrier 2014) used different machine learning algorithms to predict bike demand based on the usage record and other information about the targeting prediction time window. While the full prediction problem would be predicting bike counts at each station, the authors used machine learning to predict the bike count of the entire BSS instead. In (Dias et al. 2015), the authors used RF to classify the status of the stations only with regard to whether the station was completely full of bikes or completely empty, so users could not return a bicycle, or could not find one to rent.

In this study, we propose three major contributions to the literature. (1) Modeling bike count prediction using machine learning algorithms has not been studied well to date. (2) The univariate response models previously used to predict the number of available bikes at each station ignore the correlation between stations and might become hard to implement when applied to relatively large networks. This paper investigates the use of multivariate response models to predict the number of available bikes in the network. (3) Station neighbors, which are determined by a trip's adjacency matrix, are considered as significant predictors in the regression models.

5.2.2 Using Clustering to Explore Data Trends

Various researchers have attempted to derive insights by exploring trends through visualization techniques (Basu et al. 2003; Bar-Hillel et al. 2003; Sinkkonen et al. 2002; Demiriz et al. 1999) that are suitable for the huge and complex BSS data. For example, Froehlich et al. studied BSS patterns using 13 weeks of bicycle station usage in Barcelona (2009). They investigated the relationship between human behavior, geography, and time of day, and then tried to predict future bicycle station usage. The temporal and spatio-temporal patterns were discussed, and the results show that there are dependencies among the stations. The available bicycling data were used to cluster the docking stations. Neighboring stations were found to be highly correlated and therefore clustered in one group. Additionally, stations located on the boundaries of the bike-sharing network were found to be less active. Kaltenbrunner et al. also attempted to improve the BSS in Barcelona using docking station data (2010). Temporal and geographic mobility patterns were obtained and analyzed with the goal of detecting imbalances in the BSS. Subsequently, they used time series analysis techniques to predict the number of bicycles at a given station and time. Investigating BSS networks in terms of determined regions gives new insights to policy makers. The fact that stations in each region derived by the multivariate analysis share the same zip code implies that most of the trips were short distance, which may be influenced by the overtime fees applied when trips are longer than 30 minutes. It was also found that the most effective prediction horizon is 15 minutes. Determining prediction horizon is beneficial to policy makers and technicians to learn how to manage the BSS more responsively, and achieve better performance in prediction.

Vogel et al. attempted to derive bike activity patterns by analyzing bike share data along with geographical data (2011). Cluster analysis was used to group the bike stations with respect to pick-up and return activity. They used K-means, expectation maximization, and sequential information-bottleneck algorithms to conduct their analysis. Using the temporal activities of the stations, their results show that the bike stations could be clustered into five groups, and, thereby, average pick-up and return for each hour was given for each group. After that, the authors tried to link these five clusters with geographical information data, and found that stations in the same cluster tend to be neighbors. However, this study only clustered the bike stations based on time of day. They did not consider other temporal factors such as day of the week, month of the year, or season. Also, they used three clustering approaches, all of which clustered the bike stations based on only one objective function, the number of bikes picked up and returned at each station.

Recently, supervised clustering has been widely introduced in data mining as a powerful tool that can classify data sets effectively. It is a novel approach that enhances a clustering algorithm by using classified examples (training data) and tries to identify clusters that have high probability density with respect to a single class (Eick et al. 2004). In (Marcu 2005; Forestier 2010) several algorithms were developed using background knowledge. However, most of the proposed algorithms follow a sophisticated approach with too many parameters. Additionally, these algorithms try to achieve only one objective function, namely, increasing purity.

5.2.3 Rebalancing

BSS rebalancing is a crucial system maintenance task, as it impacts customer satisfaction and can result in a significant reduction in customer demand (Fricker and Gast 2016). Due to its importance, BSS rebalancing has been studied extensively in the literature to determine the best way to maintain the number of bikes at each station at a certain level.

In general, rebalancing can be classified as either static, dynamic, or incentivized; however, in this chapter we focus on static rebalancing only. Hernández-Pérez and Salazar-González used the branch-and-cut algorithm to solve the 1-PDTSP for problems with up to 75 vertices (2004b). In order to solve bigger instances with up to 500 vertices, they proposed two heuristic algorithms (2004a). The first heuristic is based on a greedy algorithm and K-optimality criterion, and the second heuristic is based on a branch-and-cut algorithm. Benchimol et al. proposed an integer programming model to solve the SBRP and routing problem with a single vehicle that is allowed to visit the same station more than one time (Benchimol et al. 2011). Rainer-Harbach et al. solved the 1-PDTSP for multiple trucks with different capacities that start and end the redistribution tour at separate locations with no storage space for bikes (2013). The authors decomposed the problem into two simpler problems: routing and pick-up/drop-off planning. They started by establishing the vehicle routing schedule, then used an integer programming model to find the optimal pick-up/drop-off plan associated with that tour. Schuijbroek et al. proposed a real-time and scalable solution to the 1-PDTSP by clustering stations using a maximum spanning star approximation (2013). In this solution, the vehicle is assigned to each cluster and the redistribution tour is constructed to satisfy the required service level. Li et al. used a two-step approach to solve the rebalancing problem considering multiple types of bicycles (2016). They first used a hybrid generic search to construct the truck tour, then determined the pick-up/drop-off plan using a greedy heuristic algorithm. Ho and Szeto proposed an iterated Tabu search heuristic to solve the SBRP (2014). In their formulation of the problem, they assumed that the depot was both a pick-up and drop-off node and that the depot had sufficient bikes and capacity such that the truck could stop at the depot more than once during the rebalancing procedure. Shi et al. used a modified version of a genetic algorithm (GA) to solve the 1-PDTSP for instances with between 20 and 500 vertices (2009). In their proposed GA, they adapted the pheromone idea for crossover and replaced the mutation operator with a local search procedure.

Contardo et al. used the Dantzig–Wolfe and Benders decomposition techniques to find a real-time solution for the DBRP (2012) using the most recent demand before the repositioning decisions were made. However, their approach does not work well with rapidly changing demand. Shu et al. used Poisson distribution to predict user demand for the entire planning horizon to find an improved solution for the DBRP (2013). Ghosh et al. addressed the DBRP by considering the change in demand during the day (2017). The authors decomposed the problem into two sub-problems, bike repositioning and vehicle routing, by employing Lagrangian dual decomposition and abstraction mechanisms.

5.3 DATA SET

This study used docking station data collected from August 2013 to August 2015 in the San Francisco Bay Area. The docking station data include station ID, number of bikes available, number of docks available, and time of recording. The time data include year, month, day of the month, day of the week, hour, and minute at which the docking station data were recorded. As the station data were documented every minute for 70 stations in San Francisco over 2 years, the data set contains a large number of recorded station data. Consequently, the size of the data set was reduced by sampling station data once at every quarter-hour instead of once every 1 minute (e.g., 8:00 a.m., 8:15 a.m., 8:30 a.m., 8:45 a.m., etc.) and obtaining the exact values without any smoothing process. This was done to reduce the complexity of the data and take a global view of bike availability in the entire network every 15 minutes.

During the data processing phase, we found that numerous stations were recently added to the network and others were terminated. Consequently, the data set was cleaned by eliminating any entries missing docking station data.

This reduced the number of entries from approximately 70,000 to 48,000. Each entry included the availability of bikes at the 70 stations with the associated time (month of year, day of the week, and hour of the day) and the weather information. The weather information are mean temperature, mean humidity, mean visibility, mean wind speed, precipitation, and events in a day (i.e., rainy, foggy, or sunny). These parameters were selected based on subject-matter expertise and previous related studies (Rudloff and Lackner 2014; Gallop et al. 2011), and they were found to be significant in predicting the number of available bikes at Bay Area Bike Share stations (Ashqar et al. 2016).

Moreover, we have trip data with detailed information about origin station, destination station, and time of each bike trip within the BSS during the 2 years. We used the trip data to generate the adjacency matrix of the BSS network and found the highest ten in-degree stations for station i, which were assigned as neighbors of station i. In other words, the neighbors of a station i were defined based on the number of trips originated from station j, in which $j \neq i$, and ended at station i. In the following section we will show how we used the neighbor information in building the prediction models.

5.4 BIKE PREDICTION MODELS AND RESULTS

We adapted three machine learning algorithms to build bike availability predictive models. The adapted machine learning algorithms were RF (Breiman 2001), least square boosting (LSBoost) (Friedman 2011), and partial least squares regression (PLSR) (Höskuldsson 1982; Wold 1982; Wold et al. 1984; Wold et al. 2001; Geladi and Kowalski 1986). We approached the prediction problem from two different directions. The first used RF and LSBoost to build univariate prediction models for each station in the BSS network. The second used PLSR to build one multivariate model for each zip code in the BSS network. The following subsection describes the details of both approaches and the results.

5.4.1 Univariate Models

RF and LSBoost algorithms were applied to create univariate models to predict the number of available bikes at each of the 70 stations of the Bay Area Bike Share network at time $t + \Delta$. The two algorithms were applied to investigate the effect of several variables on the prediction of the number of available bikes in each station i in the Bay Area BSS network, including the available bikes at station i at time t, the available bikes at its neighbors at the same time t, the month of the year, day of the week, time of day, and various selected weather conditions. The predictors' vector for station t at time t, denoted by X_t^i, was used in the built models to predict the log of the number of available bikes at station i at a prediction horizon time, denoted by $\log\left(y_{t+\Delta}^i\right)$, where $i = 1, 2, \ldots, 70$. The effect of different prediction horizons, Δ (range 15–120 minutes), on the performance of both algorithms was investigated by finding the mean absolute error (MAE) per station (i.e., bikes/station), which can be described as the prediction error. Moreover, as the number of generated trees by RF and LSBoost is an important parameter in implementing both algorithms, we investigated its effect by changing the number of generated trees from 20 trees to 180 trees with a 40-tree step.

As shown in Figures 5.1 and 5.2, the prediction errors of RF and LSBoost increase as the prediction horizon Δ increases. The lowest prediction error for both algorithms occurred at a 15-minute prediction horizon. Moreover, the prediction error of RF and LSBoost decreases as the number of trees increases until it reaches a point where increasing the number of trees will not significantly

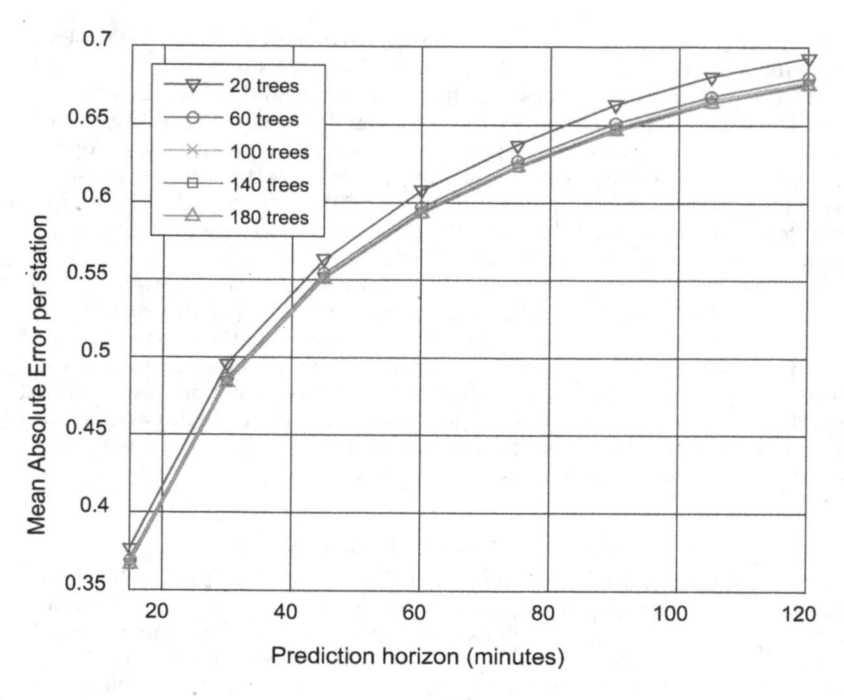

Figure 5.1 RF MAE at different prediction horizons and number of trees.

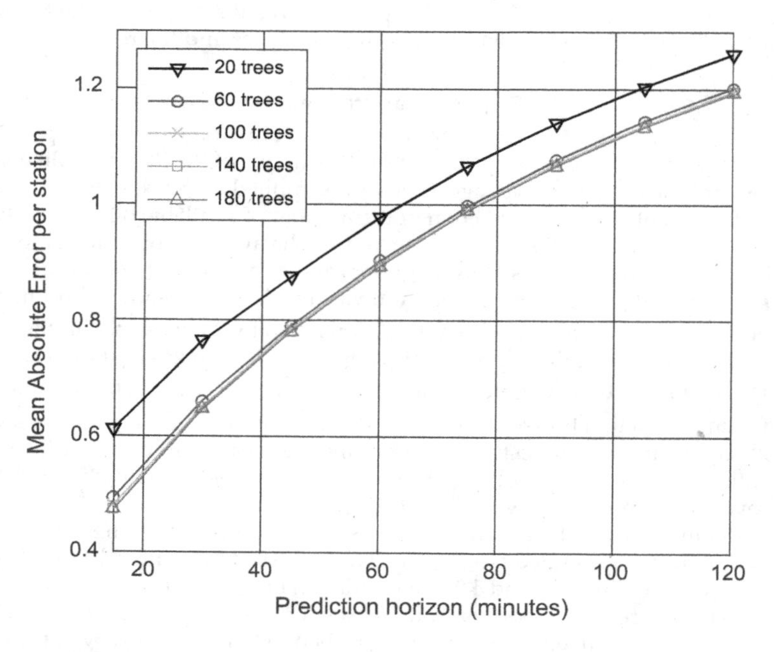

Figure 5.2 LSBoost MAE at different prediction horizons and number of trees.

improve the prediction accuracy. Figures 5.1 and 5.2 also show that a model consisting of 140 trees yields a relatively sufficient accuracy.

Comparing the two algorithms, the models produced by RF generally have a smaller prediction error than those produced by LSBoost. LSBoost is a gradient-boosting algorithm, which usually requires various regularization techniques to avoid overfitting (Ganjisaffar et al. 2011). As Figure 5.2 clearly shows, as the prediction horizon time increases, the prediction error increases (this is also clearly shown in Figure 5.3 in the next section).

5.4.2 Multivariate Models

PLSR was used as a multivariate regression to reduce the number of required prediction models for bike stations in the BSS network. When a BSS network has a relatively large number of stations, tracking all the specified models for each bike station becomes complex and time-consuming. For that reason, we examined the adjacency matrix of the Bay Area BSS network and found that the network can be divided into five regions as shown in Figure 5.4 In fact, the bike stations that resulted from the adjacency matrix in each region were found to share the same zip code. This means that the majority of bike trips occurred within the same region and very few trips went from one region to another.

Using PLSR as a regression algorithm can be used to build prediction models for multivariate response. Therefore, PLSR was applied to reduce the number of models to five, each of which is specified for one region (i.e., one zip code) to reflect the spatial correlation between stations. The input predictors' vector is X_t^z, which consists of the available bikes at all stations at region z at time t the month of the year, day of the week, time of day, and various selected weather conditions. The response's vector is $\log\left(Y_{t+\Delta}^z\right)$, which is the log of the number of available bikes at all stations' region z at a prediction horizon time Δ (range 15–120 minutes). We found that the prediction errors for PLSR were higher

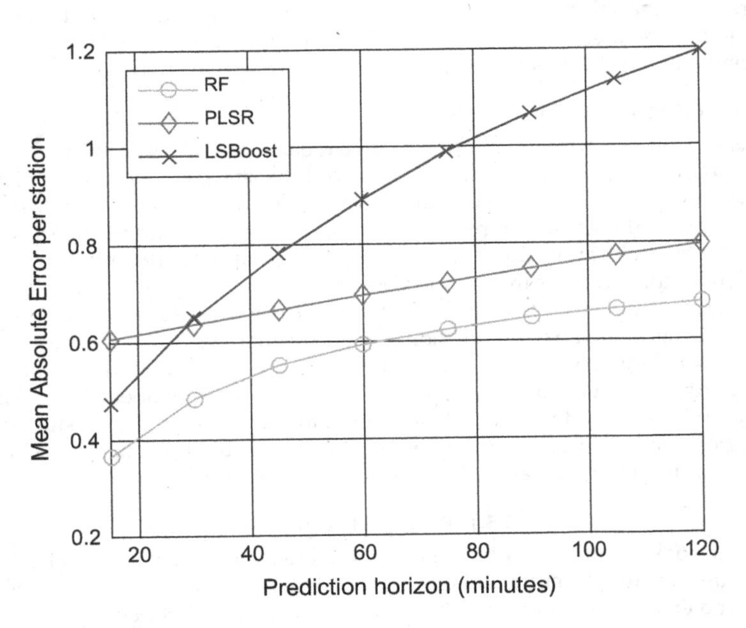

Figure 5.3 PLSR, RF, and LSBoost MAE at different prediction horizons.

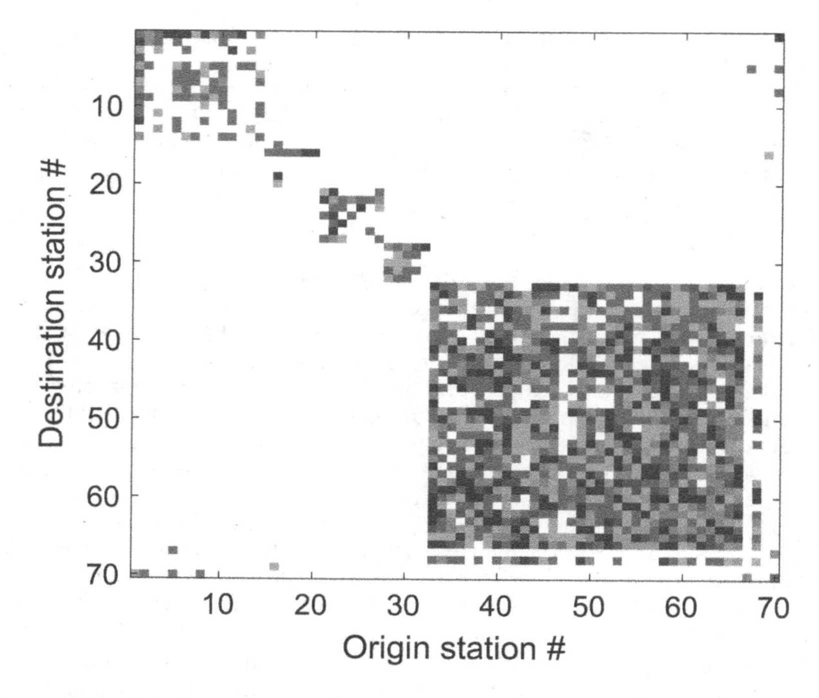

Figure 5.4 Adjacency matrix of the Bay Area Bike Share network.

than the RF and LSBoost prediction errors when $\Delta = 15$ minutes, as shown in Figure 5.3, although the prediction errors resulting from PLSR were higher than the previous results, the resulting models from PLSR are sufficient and desirable for relatively large BSS networks.

5.5 SUPERVISED CLUSTERING

In this section, we propose a new supervised clustering algorithm that has the ability to increase the purity of the cluster and the similarity of its members considering only one parameter, namely the model order (number of clusters). The proposed algorithm compromises between distortion and purity in identifying clusters within the data. We applied our algorithm to the bike availability data set without weather information. We then grouped the data into different clusters with respect to time of day and day of the week. The proposed algorithm is designed to answer questions such as what days of the week are similar, what hours of the day are similar, and what months of the year are similar. Answering these questions are really important for smartly operating the BSS. For example, we could predict the number of available bikes at each station and, thereafter, addressed when and where the system would be imbalanced by finding similar hours of the day and inspecting their centroids

5.5.1 Proposed Algorithm

Knowing some similarities in the data set is a great advantage to clustering algorithms. It can efficiently and effectively advance the outcome of the algorithm and create meaningful clusters. Accordingly, we developed a novel iterative supervised clustering algorithm. The proposed algorithm switch between partitioning that dataset using the CA algorithm (Gale and Shapley 1962) and computing the centroids of the partitions using the median. Our proposed algorithm takes advantage of the natural labeling of the data (i.e., day of the week, time of day) and models the clustering problem as a cooperative game. In this game, two disjointed sets of players join the game to identify a stable match. The first player's set consists of the centroids, and the second player's set consists of the data examples (data points). Each centroid orders the data points in its preference list based on the distance from the centroid to the data point. On the other side, each data point orders the centroids in its preference list based on the purity.

It should be noted that a cluster purity is the number of objects of the largest class in this cluster divided by size of the cluster (Xiong et al. 2009). For example, a data point that has label h will give preference to the centroid that has the highest ratio of members with label h. In other words, a data point gives higher preference to centroids when the majority of its members have the same label as its own label. After players build their preference lists, through a series of iterations, the CA tries to match between the clusters (minimizing distances) and data points (maximizing average purity) until it converges. The result of the match will be used to compute the centroids for the next iteration of the clustering algorithm. The clustering algorithm terminates when the objective function (q^t) no longer improves (threshold is less than 0.001), which involves minimizing the distance and maximizing the purity as follows:

$$\text{Purity}\left(c_i\right)^t = \frac{n_i^m}{n_i} \tag{5.1}$$

$$q^t = \frac{1}{K} \sum_{i \in \{1,\ldots,K\}} \text{purity}\left(c_i\right)^t + \sum_{i=1}^{k} \sum_{x_j \in c_i} \frac{1}{d}\left(x_j, c_i\right) \tag{5.2}$$

$$\text{The stopping criteria is } \frac{q^{t+1} - q^t}{q^t} < 0.001 \tag{5.3}$$

where:

- n_i is the number of objects in cluster i, $i \in \{1,\ldots,K\}$
- n_i^m is the number of the largest class (m) in cluster i
- q^t represents the objective function at iteration t
- c_i is the centroid of cluster i, $i \in \{1,\ldots,K\}$, $j \in \{1,\ldots,N\}$
- N is the number of data points

The following is a brief description of the proposed algorithm assuming the model order K is known:

1. Randomly choose K points as the initial centroids.
2. Form K clusters by assigning all points to the closest centroid using $L1$ norm.
3. Recompute the centroid of each cluster by computing the median.
4. Find the size of each cluster and the purity.
5. Each centroid creates its preference list of points based on $L1$ norm distance.
6. Each point creates its preference list based on purity.
7. Find the best match using the CA algorithm.
8. Recompute the purity and centroid of each cluster based on the outcome of matching using the median.

9. Compute q^t.
10. Find the size of each cluster using K-median clustering results.
11. While the stopping criteria are not satisfied, repeat steps 5–11.

To illustrate this algorithm, let us assume we have N data points and want to group them into three clusters as shown in Figure 5.5. The data points' labels are known. These labels could be any observed labels, such as the day of the week. Moreover, we assume that the true number of clusters is three. The question we want to answer is how to partition the N data points such that similar data points in terms of distance and true labels are grouped together. By effectively partitioning the N data points, we can answer questions such as which days of the week have similar bike availability across the network.

In the first step, the proposed algorithm will first randomly choose three points as centroids for the three clusters. Then, it will partition the data points based on distance to get an estimate of the size of each cluster and its purity. After that, each data point builds its preference list and each centroid builds its preference list, as shown in Figure 5.5.

In the second step, the proposed algorithm, through a series of iterations, will try to find matches between clusters and data points and provide a stable match. At the end of this iteration, all points should be matched with one of the three clusters.

After successfully matching the point with clusters, the centroid of the three clusters will be recalculated, and a new cluster size will be created. Accordingly, we repartition the N data points based on L1 norm to find the cluster size for the next iteration. The algorithm will repeat the entire process of building new preference lists, matching, and calculating new centroids. The algorithm will stop when there is no significant improvement in the objective function.

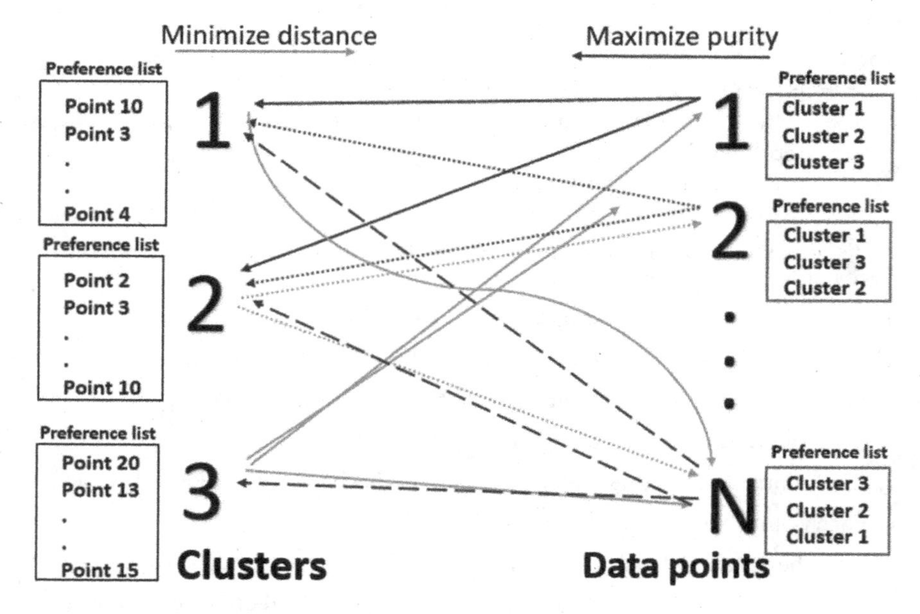

Figure 5.5 Illustrative example.

5.5.2 Model Order Selection

In cluster analysis, determining the number of clusters (K) is called model order selection. Recently, numerous cluster ensemble methods have been proposed to assist in finding the optimal K in which the result of a specific K is stable across all seeds. One is consensus clustering (CC), which subsamples the data set and clusters it, then calculates the consensus rate between all pairs of samples (Şenbabaoğlu et al. 2014). The result of this method is a similarity matrix that identifies the number of times two data points are assigned to the same cluster centroid, $k \in K$, which can be used to show the degree of stability for each model order K. One of the measures for CC that can show the cluster stability is the cumulative distribution function (CDF) against consensus rate. In the CC method, every curve represents a model order K, and the more the curve is flat, the more stable the number of clusters K is.

In our algorithm, we used the aforementioned method to determine the best K for each label. The hour label was tested for a varying value of K, and a CDF against consensus rate curve was drawn, as can be seen in Figure 5.6. Each curve represents a K and it can be seen that the curve of $K = 2$ is flatter compared to the other curves, indicating that $K = 2$ is the optimal K for the hour-of-the-day label. Consequently, we analyzed the data in more detail for $K = 2$ for the hour-of-the-day label. Similarly, we applied this technique to the day of the week, as shown in Figure 5.7, and we found that the optimal K is 3. Results for each label are discussed in detail in the following sections.

5.5.3 Day of the Week

After determining the optimal number of clusters ($K = 3$) using CC, we explored the clustering result. The results of the three clusters are presented in Figure 5.8, in which each day takes a probability of being in one of the three clusters. The three clusters are dominated by specific days: Saturdays and Sundays are clustered in the first cluster; Mondays and Fridays are clustered in the second cluster; and finally the core weekdays Tuesday, Wednesday, and Thursday are clustered in the third cluster. This pattern differs from previous research (Kaltenbrunner et al. 2010), which showed bike

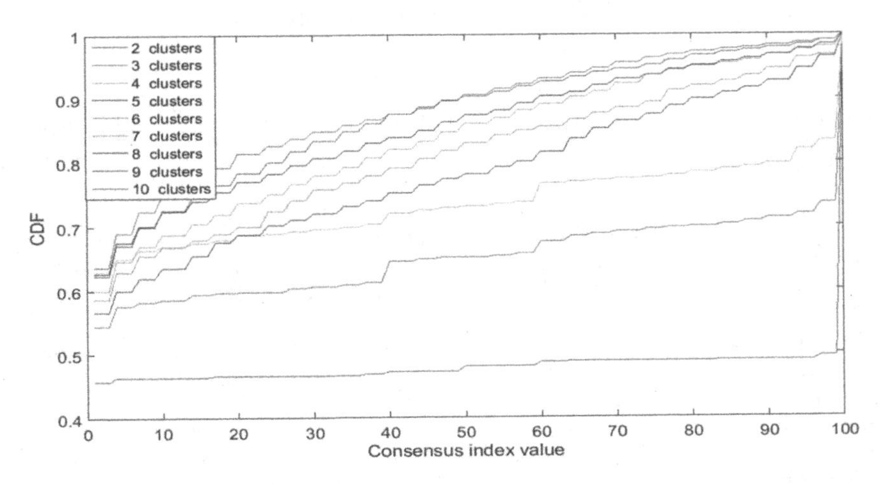

Figure 5.6 Cumulative distribution function against consensus index value for each cluster (hours) using our proposed clustering algorithm.

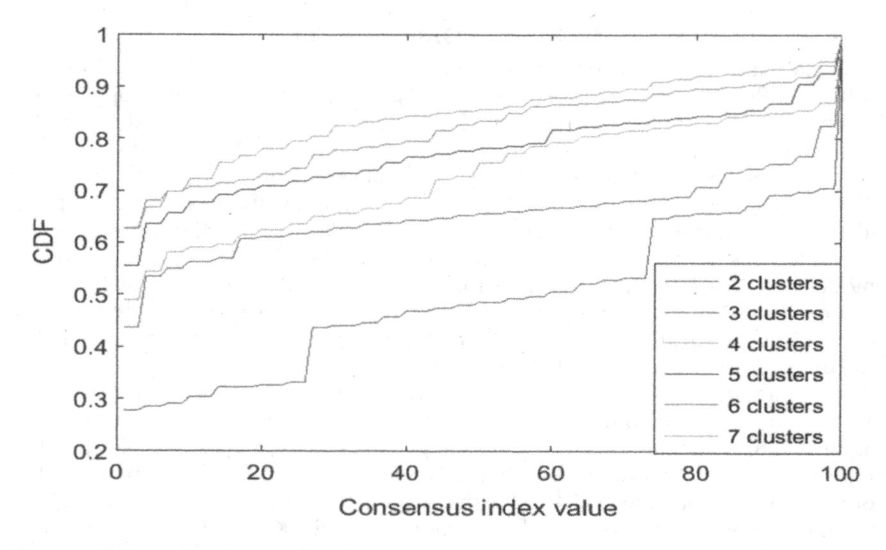

Figure 5.7 Cumulative distribution function against consensus index value for each cluster (day of week) using our proposed clustering algorithm.

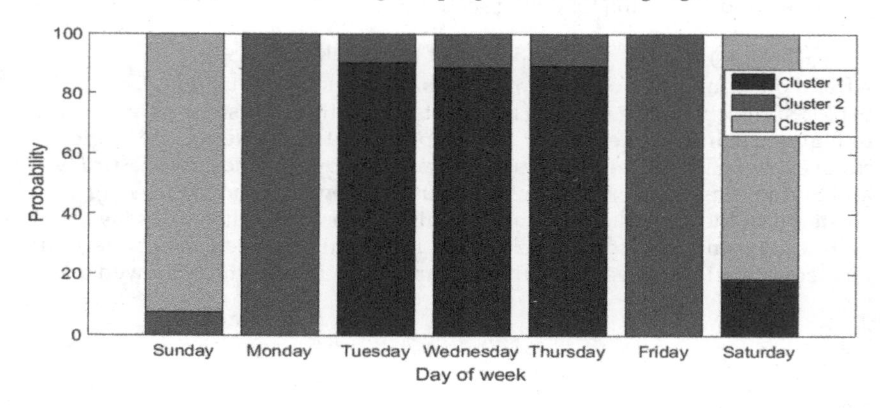

Figure 5.8 The probability of the day of week to be in one of the three clusters ($K = 3$).

patterns grouped into two clusters (weekends and weekdays). Our research shows that for this BSS the weekdays can be split into groups: (a) Mondays and Fridays, and (b) Tuesdays, Wednesdays, and Thursdays. This appears to be logical as the beginning and the end of the week are different from the rest of the weekdays.

Each cluster is associated with a pattern for the availability of bikes for each station. The patterns of the ratio of the available bikes to the station capacity for the three clusters are provided in Figure 5.9. It should be noted that the pattern represents the centroid of the cluster (the median of the cluster) so that the exact values of the pattern are not shown here. The pattern can serve as an indication of when the station is more likely to be full or empty on any specific day of the week. Three observations can be drawn from Figure 5.9. First, the

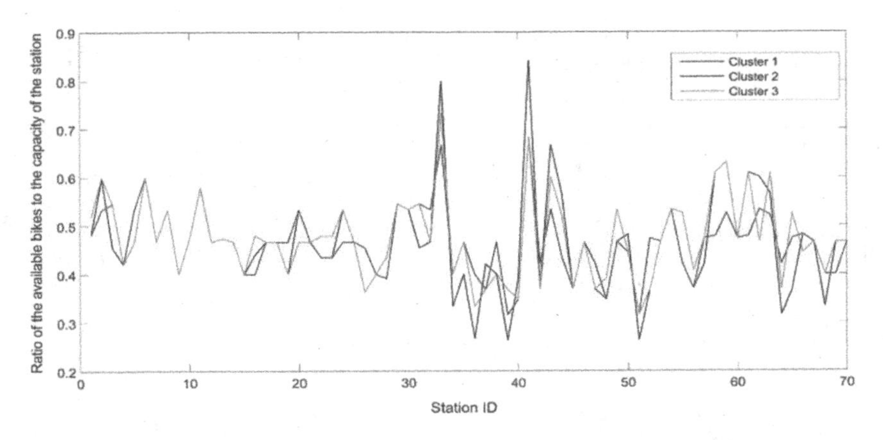

Figure 5.9 The ratio of the availabile bikes to the capacity of the station for the three clusters at station in the network.

pattern of the three clusters shows fluctuations in the bike activities. None of the days of the week has the highest activity for the entire network, which depends on both spatial and temporal factors. Second, several stations appear to be more likely empty or full on either weekdays or weekends. The difference in demand between the three clusters appears clearly for some stations, but not others. For example, station 31 (Cowper at University), located in Palo Alto, and station 41 (Embarcadero at Folsom), located in San Francisco, are rarely used by bikers, especially on Tuesdays, Wednesdays, and Thursdays (i.e., bikers might not be able to drop their bikes at these stations on these days as they are almost full). In contrast, station 36 (Washington at Kearney) and station 39 (Spear at Folsom) are highly used by bikers, and thus are more likely to be empty. Consequently, it might be wise to consider either relocating or changing the size of these stations. Surprisingly, when analyzing the location of station 41, we found that station 39 (Spear at Folsom), which is highly used by bikers, is very close (only one block away); thus, it might be better to move bikes between them overnight, especially on Tuesday, Wednesday, and Thursday, as the largest differences in usage occur on these days. Another suggestion is to incentivize bikers to drop-off their bikes at station 39 instead of station 41 or pick-up their bikes from station 41 instead of station 39. Third, the three clusters have the same pattern for stations 6 to 15, 67, and 70. In other words, all the days of the week have the same bike activities. When examining their locations, we found that all of these stations are located in San Jose, so we can imply that the bike activities in San Jose are approximately similar during the entire week. That could be caused by the fact that both tourists and workers use the bikes for their trips, or that the traffic patterns in San Jose are the same during the entire week.

The bike activities for cluster 1 (Tuesday, Wednesday, and Thursday) and cluster 3 (Saturday and Sunday) are similar for some stations in the network. That can be seen in stations 58 and 59 (San Francisco Caltrain 2, 330 Townsend and San Francisco Caltrain–Townsend at 4th respectively). When taking a closer look at the location of these two stations, we found that these two stations are located close to the Caltrain station. Consequently, the similarity between these two clusters can be linked to the train timetable.

5.5.4 Hour of the Day

The station data were clustered with respect to the hour-of-day label without taking into account the-day-of-week label. In other words, we tried to find the peak and non-peak hours for any day of week. Only the station data at the beginning of each hour were considered (e.g., 8:00 a.m., 9:00 a.m., 10:00 a.m., etc.). The optimal number of clusters was found to be 2 ($K = 2$) as shown in Section 5.5.2. The analysis of the data reveals that the two clusters are peak (cluster 2) and non-peak (cluster 1) hours, thus confirming the previous research. The results of the clustering are given in Figures 5.10 and 5.11, which show the probability of an hour being in one of the two clusters and the pattern of each cluster. It can be concluded from Figure 5.11 that when the patterns of the two clusters are lined up, the bike activity in the peak and non-peak hours is the same. For the sake of space, we will not go in detail with regard to the hour label.

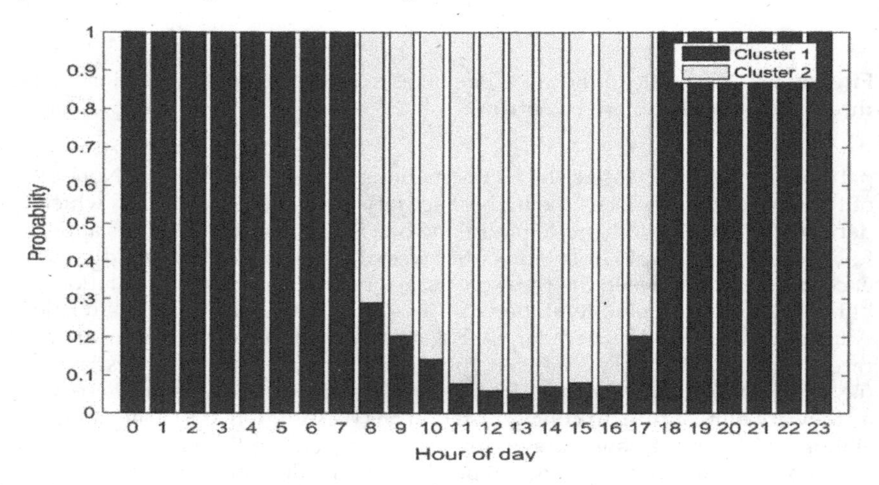

Figure 5.10 Probability of hour to be in one of the two clusters ($K = 2$).

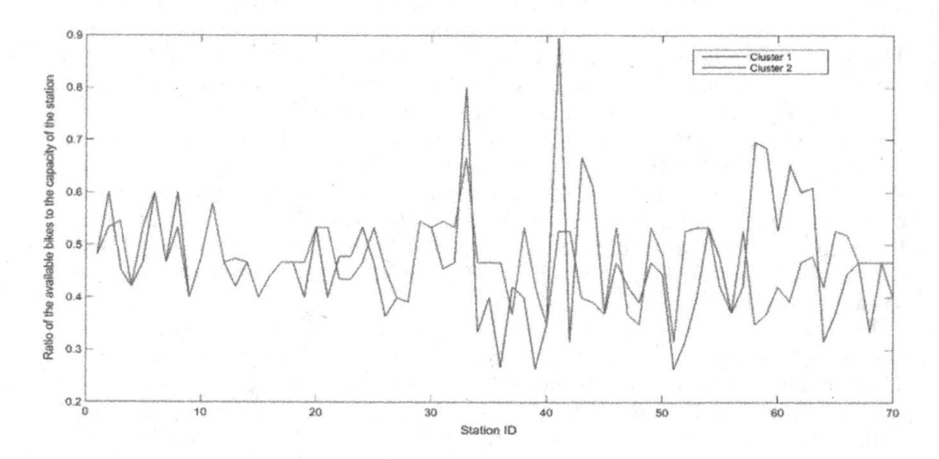

Figure 5.11 The ratio of the available bikes to the capacity of the station for the two clusters at station in the network.

5.6 REBALANCING

In this section, we propose an algorithm for solving the SBRP. The proposed algorithm consists of two stages: tour construction and tour improvement. In the tour construction stage, we model the SBRP using two sets of disjointed players. Each player constructs their own set of preferences for the opposite set of players, then the deferred acceptance algorithm (Gale and Shapley 1962) is used to find stable matches between the two sets. In the tour improvement stage, a 2-opt algorithm is used to search the neighborhood of the constructed tour to find a better tour. The proposed algorithm has the advantage of being able to easily consider a change in the constraints by restricting each player's preference list changes to the new constraints. Accordingly, this algorithm can be easily used for dynamic rebalancing provided the demand predictions are known. Moreover, the matching algorithm and local search algorithm run in polynomial time, which means the proposed algorithm can solve large instances and is suitable for real-time application.

5.6.1 Problem Statement

In order to state the problem, we must first define the term "NotSpot," which is adopted from (Goodyear 2014). A NotSpot is a bike station that is empty or full, rendering it useless to people who want to pick-up or return a bike. A bike station that has a number of bikes less than a certain threshold t_{Empty}, or greater than a threshold t_{Full}, is considered a NotSpot as well. The thresholds t_{Empty} and t_{Full} can be estimated for each station from its historical data set. However, the estimation of these thresholds is beyond the scope of this research.

Given a fully connected graph $G = (V, E)$, where V is the depot and the set of M NotSpots in the bike system, and E consists of all the links connecting the vertices in the graph. Each node i in the graph is assigned an integer β_i that equals the number of needed/pick-up bikes. β_i is positive if it represents the number of bikes that need to be taken at a NotSpot, making station i a pick-up station. β_i is negative if it represents the number of needed bikes at a NotSpot, making station i a drop-off station. Each link j is assigned a cost γ_j.

The research question we need to address is, given a list of NotSpots, the distances between each pair of NotSpots, the demand of the NotSpots, and only one truck with maximum capacity Q, what is the optimum tour for the truck to rebalance the NotSpot stations?

The tour is optimum in the sense that the objective function $\sum_{j\in E}\gamma_j + \sum_{i\in V} R_i$ is minimized subject to $\begin{cases} AC_i > 0 & \text{if } \beta_i > 0 \\ Q - AC_i \geq 0 & \text{if } \beta_i < 0 \end{cases}$ where R_i, Q, and AC_i are the residual at NotSpot i, the maximum capacity, and the available bike spots on the truck before serving NotSpot i, respectively. This constraint guarantees that no full truck arrives at a pick-up NotSpot station and no empty truck arrives at a drop-off NotSpot station.

Recall that this problem is 1-PDTSP. In the context of the BSS, we assume we have only one type of bike, so it is a one-commodity problem. Moreover, we assume that the tour starts from the depot and ends at the depot. In this problem, the truck could start the tour from the depot with any number of bikes less than or equal to its maximum capacity.

5.7 PROPOSED ALGORITHM

The proposed algorithm consists of two phases. The first phase, which is the main phase, is the tour construction. Tour construction is based on applying the deferred acceptance algorithm M times to match two disjoint sets: the partial tour and the NotSpot stations. The matching is done such that at the end of the construction phase, each constructed tour includes all NotSpot stations and each NotSpot station appears only once in the tour. The second phase uses the 2-opt algorithm to improve the constructed tours and then selects the best tour as the solution to the given problem. In the following subsections, we will describe the two phases in more detail.

5.7.1 Tour Construction Using the Deferred Acceptance Algorithm

The ultimate goal of the proposed tour construction algorithm is constructing optimized M tours. Each tour consists of M NotSpots in addition to the depot and is a Hamiltonian path. We modeled the problem of the tour construction as a cooperative game between two disjoint sets of players. The first player's set consists of M partial tours, where each tour is represented by the current load H_{i-1} of the truck after serving the last NotSpot in the partial tour. The objective of each partial tour is finding the next NotSpot i to serve. Thus, each partial tour builds an ascending list based on

$$(\alpha Q - H_i)^2 + B \times D \tag{5.4}$$

where:

- Q is the maximum truck capacity
- α is a constant between 0.1 and 1.0
- H_i is the number of available bikes on the truck after serving NotSpot i
- B is a constant $\in [0, 1]$
- D is the distance in meters between the last station in the tour and NotSpot i

The partial tour's preference list contains only the NotSpots that are not shown in its current path, and is ordered such that the most preferred NotSpot has the minimum value of Equation 5.4.

The other player's set consists of M NotSpots. Each NotSpot i builds its preference list such that it lists only partial tours that do not yet include i. The partial tours preference list is ordered such that the tour that minimizes the residual at NotSpot i is at the top. The last partial tour on the list is the tour with the worst residual at NotSpot i. In other words, the top tour in this player's list satisfies the station demand the most and the last tour satisfies the demand the least.

Once the preference list is built for each player, we find a stable match between the partial tours and the NotSpots using the deferred acceptance algorithm. After matching, each partial trip gets expanded by stacking its matched NotSpot to its end. Moreover, the current capacity of the truck assumed to make the tour is updated. Finally, we determine whether or not the partial trip includes all the NotSpots. If all NotSpots are *not* included, then each player's preference list is rebuilt and another game is played. If all NotSpots *are* included, the algorithm is stopped. The pseudocode of the proposed algorithm is shown in Table 5.1.

Table 5.1: Pseudocode of the Proposed Algorithm for Tour Construction

1. Construct M partial tours, each consisting of the depot and one station of M NotSpots.
2. Update the truck capacity for each partial tour.
3. For $f = 1 : M - 1$
   a. Each NotSpot i builds its preference of partial tour $pt \in \{1, 2, \ldots, M\}$ based on the residual at NotSpot i if it was stacked to partial tour pt.
   b. Each tour builds its station preference list using $(\alpha Q - H)^2 + B \times D$.
   c. Run the deferred acceptance algorithm to match the stations with the partial tours.
   d. Update the current available bikes on each tour truck after expanding the partial tour.
4. Return M constructed tours, with each tour being a Hamiltonian path.

Table 5.2: Pseudocode of the Proposed Algorithm for Tour Improvement

1. For each tour constructed by the deferred acceptance algorithm, calculate the total tour cost using Equation 5.2.
2. For $f = 1 : M$
   a. Find all possible 2-opts of the constructed tour f.
   b. For each 2-opt of the constructed tour f, calculate the total tour cost and find the best modified tour.
   c. If the best 2-opt tour has less cost than the original tour constructed by the deferred acceptance algorithm, replace the original tour f with its modified 2-opt tour.
3. The solution is the best trip out of the M tours.

5.7.2 Tour Improvement Using 2-Opt Local Search Algorithm

At the end of the tour construction phase, we will get M different tours, with each tour consisting of M NotSpots. In the tour improvement phase, we evaluate the cost of each of the M constructed tours using Equation 5.5.

$$B \times \sum_{j \in E} D_j + \sum_{i \in V} R_i^2 \tag{5.5}$$

where:

D_j is the length of link

j in meters

R_i is the residual at NotSpot i

B is the same constant used in Equation 5.4

For each constructed tour, we find all possible 2-opt algorithms and evaluate each. If the best 2-opt tour has a lower cost than its original constructed tour, then the constructed tour is replaced by its 2-opt. After finishing the local search for all M constructed tours, we choose the tour that has the lowest cost as a solution to the rebalancing problem. The pseudocode of the tour improvement algorithm is shown in Table 5.2.

5.7.3 Tour Construction Example

In order to illustrate the proposed algorithm for tour construction, we consider a fully connected undirected graph consisting of three NotSpot stations and depot. For simplicity, we assume the vertices form a square with each side 1000m long, with the exception of the link between $S1$ and $S2$, which has a length of 1,200 m, as shown in Figure 5.12. We set B, Q, and α equal to 0.002, 10, and 0.5, respectively. Moreover, we assume the truck has four bikes when it leaves the depot.

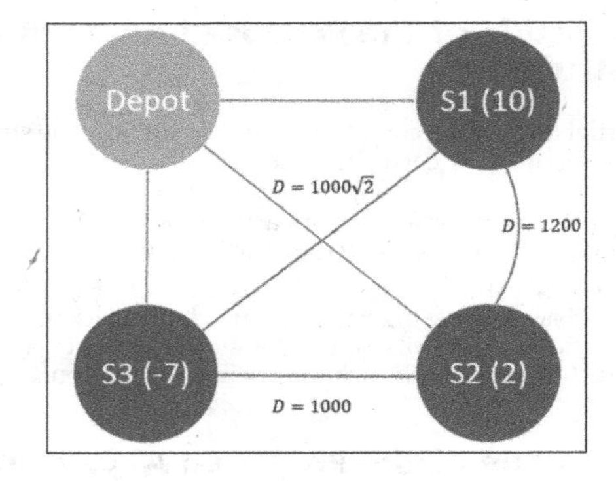

Figure 5.12 Illustrative example graph where the number in parentheses is the NotSpot demand.

The tour construction algorithm consists of the following steps:

1. Construct three partial tours, where each consists of the depot and one of the three NotSpot stations, then update the truck capacity H of each tour as shown in Equation 5.6.

$$\begin{aligned} &\text{Depot} \rightarrow S1 \quad (H_{S1} = 10) \\ &\text{Depot} \rightarrow S2 \quad (H_{S2} = 6) \\ &\text{Depot} \rightarrow S3 \quad (H_{S3} = 0) \end{aligned} \tag{5.6}$$

2. Set the preference list for each partial tour using Equation 5.4. The parameter α is set to 0.5 to give preference to pick-up or drop-off stations that bring the truck to its half capacity.
3. Set the preference list for each NotSpot such that the partial tours that minimize the residual at the NotSpot are on the top of the list as shown in Figure 5.13.

The first set of players	The second set of players
Depot→S1 preference list	**S1 preference list**
S3 $(10/2-(10-7))^2+.002*1000\sqrt{2}$	**Depot→S3**$(R_{S1}=0)$
S2 $(10/2-(10))^2+.002*1200$	**Depot→S2**$(R_{S1}=6)$
Depot→S2 preference list	**S2 preference list**
S3 $(10/2-(0))^2+.002*1000$	**Depot→S3**$(R_{S2}=0)$
S1 $(10/2-(10))^2+.002*1200$	**Depot→S1**$(R_{S2}=2)$
Depot→S3 preference list	**S3 preference list**
S2 $(10/2-(2))^2+.002*1000$	**Depot→S1**$(R_{S3}=0)$
S1 $(10/2-(10))^2+.002*1000\sqrt{2}$	**Depot→S2**$(R_{S3}=1)$

Figure 5.13 The preference list for the two sets of players.

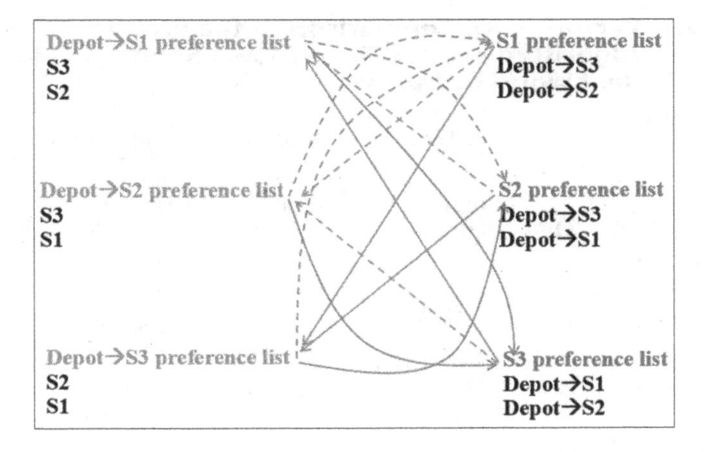

Figure 5.14 When the partial tours make offers, the first and second partial tours choose S3 and the third partial tour chooses S2. S3 accepts the first partial tour (S3's first choice) and S2 accepts the second partial tour. In a second stage, the second partial tour makes an offer to S1, which accepts, and the game results in a stable match.

4. Match the NotSpots with the partial tours using the deferred acceptance algorithm. Based on the matching outcome, the three partial tours are expanded as shown in Figure 5.14 and Equation 5.7.

$$\begin{aligned} &\text{Depot} \rightarrow S1 \rightarrow S3 \\ &\text{Depot} \rightarrow S2 \rightarrow S1 \\ &\text{Depot} \rightarrow S3 \rightarrow S2 \end{aligned} \tag{5.7}$$

5. Update the number of currently available bikes for each truck following the expanded partial tours.
6. Check the number of stations in the partial tour; if the number is less than 3, then go to step 2.

5.8 RESULTS

The proposed algorithm was coded in MATLAB® 2015b. One advantage of the proposed algorithm is that it has only two hyperparameters: α and B. In the first set of experiments, we set α and B equal to 0.5 and 0.2, respectively. The proposed algorithm was tested using real data collected by Dell'Amico et al. from La Spezia, Italy; Wisconsin, United States; and Ottawa, Canada (2014). For the sake of comparison with other heuristic algorithms, we chose the same instances used to test the ant colony optimization (ACO) algorithm proposed in (Bortner et al. 2015). The ACO algorithm is inspired by the behavior of real ants. Real ants can find the shortest path from the nest to food sources by communicating information among individuals regarding a path by depositing a certain amount of pheromone while walking. Each searching ant probabilistically prefers to follow the path with the highest pheromone levels rather than paths with lower pheromone levels. Readers who are interested in the implementation of the ACO for rebalancing BSSs should consult (Bortner et al. 2015).

Table 5.3: Performance Comparison of the Proposed Algorithm and the Best-Known Solution Value for Real World Instances

			ACO (Bortner et al. 2015)		Proposed Algorithm	
	N	Q	Total Distance (m)	Total Residual	Total Distance (m)	Total Residual
LaSpezia	20	30	21,518	1	21,475	1
		20	23,488	1	21,475	1
		10	23,908	1	23,529	1
Ottawa	21	30	17,178	0	17,166	0
		20	17,178	0	17,166	0
		10	17,604	2	17,166	2
Madison	28	30	34,029	18	35,877	8
		20	34,706	8	34,420	8
		10	38,677	8	37,271	8

Table 5.3 shows the comparison between our proposed algorithm and the ACO. The results show that our proposed algorithm returns a better solution for most of the instances. The instances used in the comparison to ACO have an absolute total demand close to zero, which ensures that the system had enough bikes for balancing the BSS.

We also tested the algorithm using other real instances where the absolute total demand was not close to zero. The instances were collected from Denver, United States; Dublin, Ireland; Ciudad de Mexico, Mexico; and Minneapolis, United States, and they have absolute total demands of 35, 64, 87, and 92, respectively. In solving this set of instances, we varied α from 0.1 to 1.0 and determined the best solution.

As shown in Table 5.4, the proposed algorithm returned solution values close to the best-known solutions in a number of instances. Moreover, our algorithm returned solution values better than the best-known solution values for other instances, especially when the maximum capacity of the truck was small. We should note that the residual shown in the table is due to the imbalance in the total demand. For example, the Minneapolis instance had a total demand of 92, which means that the number of pick-up bikes was more than the number of drop-off bikes by 92 bikes. In this case, when the truck's maximum capacity was 10, we found that the best solution had a residual of 82 bikes, with 10 bikes loaded on the truck at the end of the tour.

5.8.1 San Francesco Bay Area Instances

In order to check the performance of the proposed algorithm in terms of residuals, distance and running time, we created real instances. The bay area network consists of 70 stations, however, we found that the stations are geographically distant. We decided to divide the stations into three groups based on location. These groups are located around San Francisco, Palo Alto/Mountain View and San Jose. The rebalancing problem is more apparent in the first group of stations simply because the city of San Francisco is the most urban area in the bay area. The other two groups are located in suburban areas. Therefore, we chose to apply the rebalancing technique on the first group. This group consists of 35 stations. Besides, we need to specify a depot point where the rebalancing truck will start

Table 5.4: Performance Comparison of the Proposed Algorithm and the Best-Known Solution Value for Real World Instances

			Best-Known Solution (Erdoğan et al. 2015)	Proposed Algorithm		
	N	Q	Total Distance	Total Distance	Total Residual	α
Denver	51	30	51,583	53,740	5	0.7
		20	53,369	57,116	15	0.9
		10	67,025	63,281	25	0.4
Dublin	45	30	33,548	41,274	34	0.7
		20	39,786	40,489	44	1.0
		11	54,392	48,172	53	1.0
Ciudad de Mexico	90	30	88,227	79,234	57	0.2
		20	116,418	88,213	67	0.3
		17	109,573	91,188	70	0.3
Minneapolis	116	30	137,843	186,716	62	1.0
		20	186,449	237,856	72	0.4
		10	298,886	294,405	82	0.6

from and end at. We choose the depot to be the point of minimum total distance to all other 35 stations. Then we used Google Directions API to get the distance matrix between each pair of stations and between stations and depot and vice versa as shown in Figure 5.15.

In order to define the demand for the studied stations, the following has been done;

1. We visually inspected bike counts at each station trying to find a specific time in which the bike count is unlikely to change. It was found that the bike count tends to stay constant from 11:00 p.m to 12:00 a.m.
2. We randomly selected 42 days to represent different demand patterns then for each day we create a vector of bike counts at each station of the 35 stations at 11:00 p.m.
3. We assumed the goal of the rebalancing is to keep each station half full. So we computed the demand for each instance by subtracting the vector of stations' half capacity from the corresponding bike count vector obtained in step 2.

All the 42 instances have negative sum of demand which means network needs bikes. So that, we used the proposed algorithm to solve the 43 instances at different $Q \in \{10,20,30\}$ and $H \in \{10,20,30\}$ such that $H \le Q$.

The solutions to the different instances presented in this section were performed using an Optiplex 9020 Dell desktop, which has 8 GB RAM and an Intel® Core™ i7-4790 CPU @ 3.60 GHz. Figure 5.16 shows the histogram of the running time for the different combinations of Q and H. As shown in Figure 5.16, the average running time of the algorithm to find a solution is almost 60 seconds, which is sufficient. We should highlight that this average running time could be reduced to almost one tenth (6 seconds) if we use 10 cores

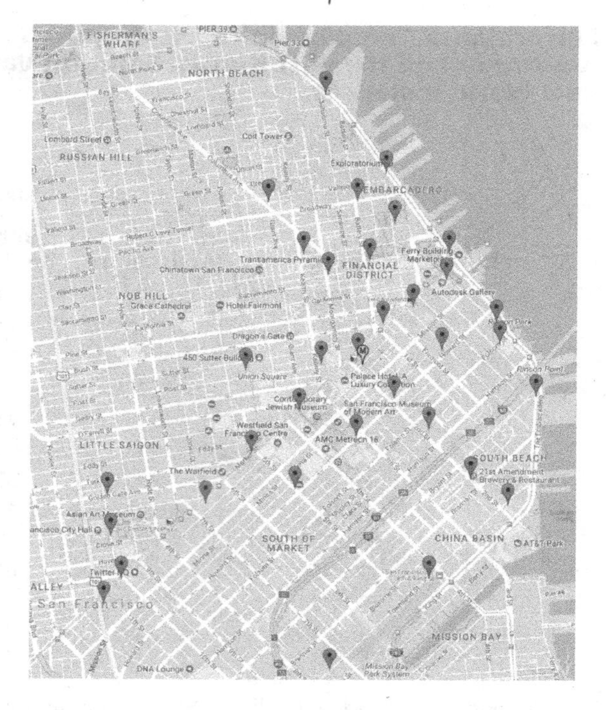

Figure 5.15 The locations of real instances' stations and depot (source: google maps).

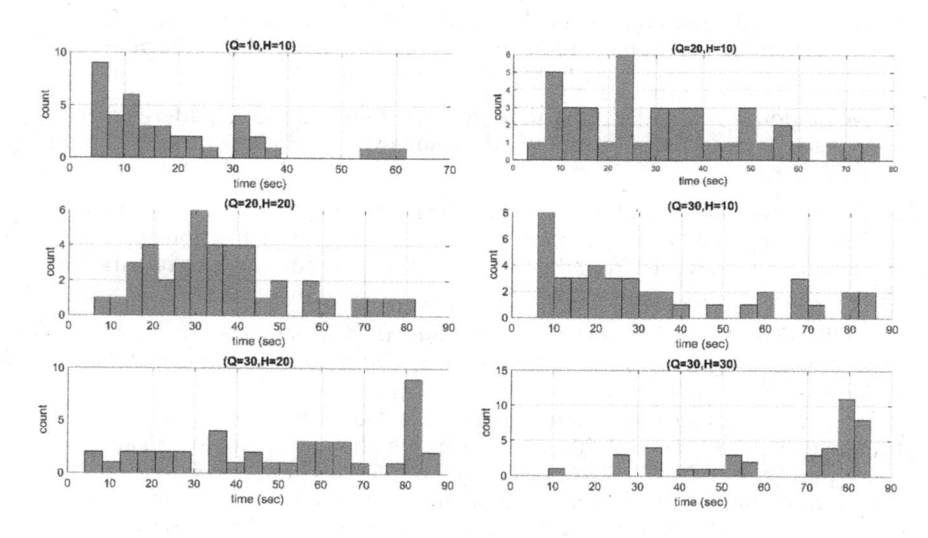

Figure 5.16 The running time of the proposed algorithm to solve the real instances.

and each core finds the solution at a given α independently and then selects the best of the 10 solutions.

In order to measure the performance of the proposed algorithm in terms of residuals, we introduced the term residual gap (RG) in Equation 5.8.

$$RG = \begin{cases} \sum_{i=1}^{35} \text{abs}(R_i) + \min\left(0, \sum_{i=1}^{i=35} \beta_i + H\right) & \text{if } \sum_{i=1}^{i=35} \beta_i \leq 0 \\ \sum_{i=1}^{35} \text{abs}(R_i) - \left(\sum_{i=1}^{i=35} \beta_i + H\right) & \text{otherwise} \end{cases} \tag{5.8}$$

where β_i, R_i and H are the demand at NotSpot i, the residual at NotSpot i and the number of available bikes on the truck after leaving the depot respectively. As shown in Figure 5.17, RG is zero for most of the instances. This does not mean most of the instances have zero residual but means that algorithm found the best solution in terms of residual. For example, assume that an instance has sum of demand equals −45 and the truck leaves the depot carrying 20 bike, this mean that the algorithm will not achieve sum of absolute residual less than 25. So if the algorithm found a route which has a residual equals 25 bikes, this route will have zero RG.

The last performance measure is the length (distance) of the trip in kilometer. We noticed that the distance depends on the demand pattern so it does not make sense to report the average distance of the different trips. In order to show how good is the resulted trips, we randomly permuted each trip and assumed that new trips satisfy the demand then compute their distances. Figure 5.18 shows how good is our solution compared to the random permutation with regard to the distance, still the resulted trips form the proposed algorithm satisfy the demand constraints.

We also visualize the solution for of one of the instances at $Q = 30$ and $H = 20$ in Figure 5.19 where each station's location is marked by a number in a blue circle. This number is the order of this station in the trip. The demand of each station is written beside the blue circle. The trip starts from the depot and ends

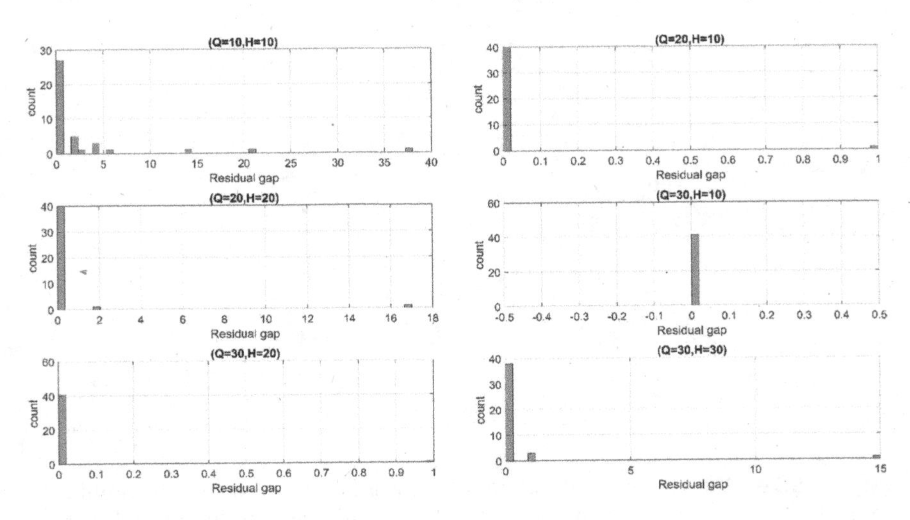

Figure 5.17 The residual gap of the proposed algorithm at different H and Q.

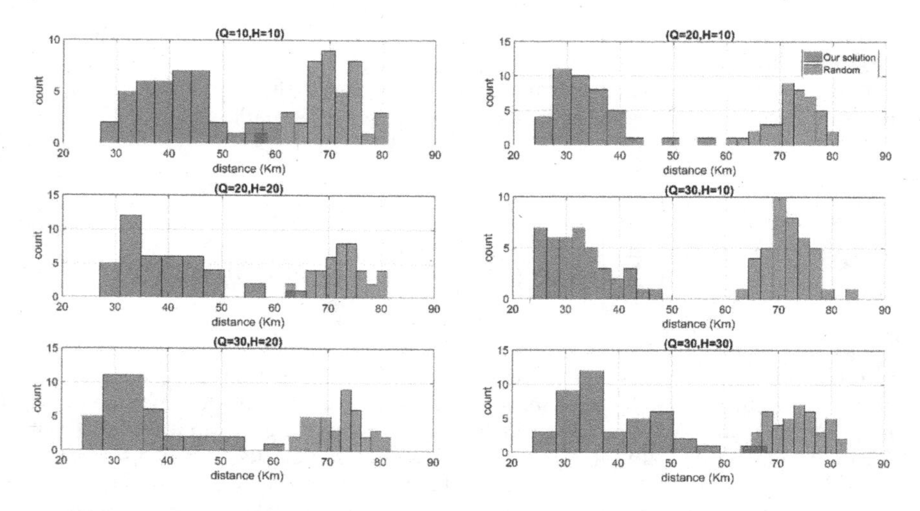

Figure 5.18 Comparison of the distance of trips using the proposed algorithm and distance of the random permutation of them.

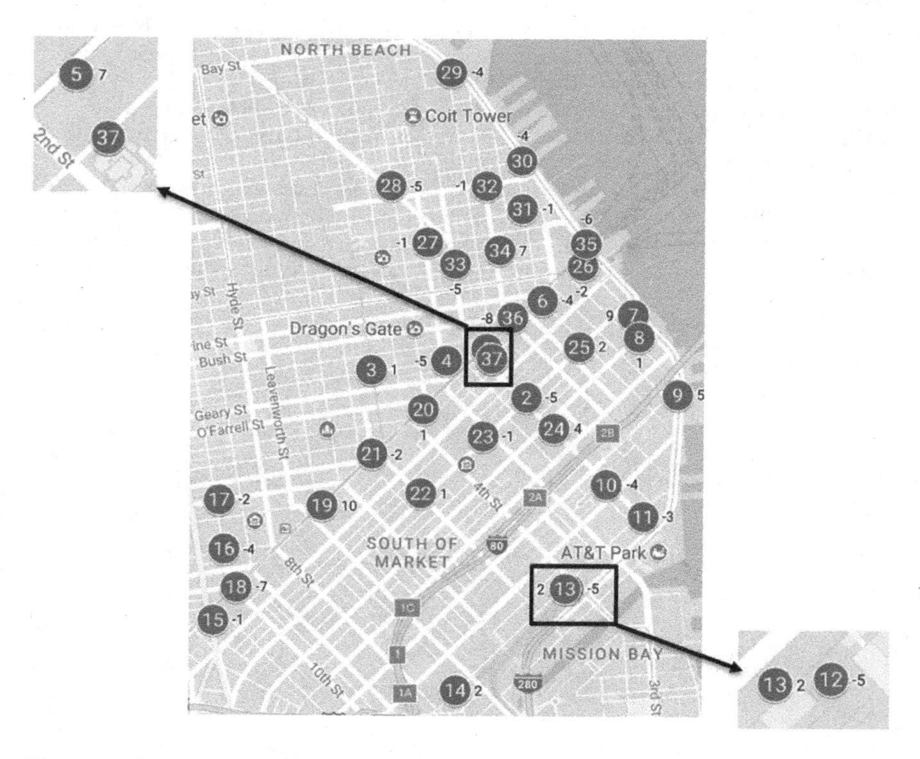

Figure 5.19 Visualization of the solution for one instance where the number in the blue circle is the station visit order and the number beside the circle is the demand of the station.

at the depot so that the circle number 37 is the depot location. The reader can verify the solution for the truck leaving the depot (node number 37) with 20 bikes and heading toward node number 2 (second station in the trip after the depot), by moving to station in the order in the blue circle and sum the demand beside the circle. Finally, we should note that truck moves from station to a station in the neighborhood and does not cross the whole network back and forth to satisfy the stations' demand.

5.9 CONCLUSIONS

In this chapter we defined and showed how SBSSs are important for connecting different transportation networks in a smart city in order to establish smart transportation. Moreover, the chapter makes three contributions related to building a toolbox of models and algorithms to convert a current BSS into an SBSS.

The first contribution is adapting state-of-the-art machine learning algorithms to model the number of available bikes. We applied these algorithms to the Bay Area Bike Share stations in San Francisco. We tried two approaches: using univariate regression algorithms, RF and LSBoost, and using a multivariate regression algorithm, PLSR. The univariate models were used to model the number of available bikes at each station. RF, with an MAE of 0.37 bikes/station, outperformed LSBoost, with an MAE of 0.58 bikes/station. On the other hand, the multivariate model, PLSR, was applied to model available bikes at spatially correlated stations of each region obtained from the trip's adjacency matrix. Results clearly show that the univariate models produced lower error predictions compared to the multivariate model, in which the MAE was approximately 0.6 bikes/station. However, the multivariate model results might be acceptable and reasonable when modeling the number of available bikes in BSS networks with a relatively large number of stations. Investigating BSS networks in terms of determined regions gives new insights to policy makers. The fact that stations in each region derived by the multivariate analysis share the same ZIP code implies that most of the trips were short distance trips. This may be influenced by the overtime fees applied when trips are longer than 30 minutes. With the most effective prediction horizon being 15 minutes, determining prediction horizon is beneficial to policy makers and technicians to learn how to manage BSSs more responsively, and achieve better prediction performance.

The second contribution is proposing a new supervised clustering algorithm that potentially will assist agencies and researchers to anticipate bike availability at stations with respect to a time event. We tested the proposed algorithm on a BSS in the San Francisco Bay Area. The proposed algorithm clusters bike availability data at 15-minute intervals across the network and finds the similarity between them according to day of the week and hour of the day. Subsequently, it provides an expected pattern of bike usage for each cluster. The algorithm provides insight into the usage patterns of the San Francisco Bay BSS that operators can use to anticipate imbalances in the system and plan accordingly. Moreover, the clustering results show that the days of the week can be grouped into three clusters, one for weekends and the other two for weekdays. The time of day is clustered into two groups, peak and off-peak hours. Given that each cluster has an associated pattern of bike availability, a prediction can be made to identify the imbalance in the system for each day of the week and each hour of the day. An exploratory spatio-temporal analysis was conducted, leading to different suggestions on how to rebalance the system with minimum cost and effort, thus making the network a more-effective component of the smart transportation system in smart city.

The third contribution is proposing a novel heuristic algorithm, based on the deferred acceptance algorithm, to solve the static rebalancing problem for SBSSs. The proposed algorithm consists of two stages. The first is the tour construction stage, which begins by assuming that each NotSpot and the depot form a partial tour. Each partial tour is then grown by matching the partial tours with the NotSpots. The matched pairs of partial tours and NotSpots are subsequently stacked, and the matching game is repeated until each partial tour is composed of all NotSpots. The second stage takes the constructed tours and uses the 2-opt algorithm to search its neighborhood for a better solution. We compared the proposed algorithm to the ACO, and the results show that our proposed algorithm outperforms the ACO in most instances. Moreover, we solved medium-size instances with up to 116 NotSpots, then compared our solution to the best-known solution. We found that our solution was close to the best-known solution for some instances and better than the best-known solution for other instances.

ACKNOWLEDGMENTS

This effort was partially funded by the Mid-Atlantic Transportation Sustainability University Transportation Center, the Urban Mobility and Equity Center, and the National Science Foundation UrbComp project.

REFERENCES

H. I. Ashqar, M. Elhenawy, and H. A. Rakha, "Modeling Bike Counts in a Bike-Sharing System Considering the Effect of Weather Conditions," unpublished, 2016.

H. I. Ashqar, M. Elhenawy, M. H. Almannaa, A. Ghanem, H. A. Rakha, and L. House, "Modeling bike availability in a bike-sharing system using machine learning," in *5th IEEE International Conference on Models and Technologies for Intelligent Transportation Systems (MT-ITS)*, 2017, pp. 374–378.

A. Bar-Hillel, T. Hertz, N. Shental, and D. Weinshall, "Learning distance functions using equivalence relations," in *Proceedings of the 20th International Conference on Machine Learning (ICML-03)*, 2003, pp. 11–18.

S. Basu, M. Bilenko, and R. J. Mooney, "Comparing and unifying search-based and similarity-based approaches to semi-supervised clustering," in *Proceedings of the ICML-2003 Workshop on the Continuum from Labeled to Unlabeled Data in Machine Learning and Data Mining*, 2003, pp. 42–49.

Bay Area Bike Share, Introducing *Bay Area Bike Share, Your New Regional Transit System*. 2016. Available: www.bayareabikeshare.com/faq#BikeShare101.

M. Benchimol, P. Benchimol, B. Chappert, A. De La Taille, F. Laroche, F. Meunier, et al., "Balancing the stations of a self service 'bike hire' system," *RAIRO-Operations Research*, vol. 45, pp. 37–61, 2011.

C. W. Bortner, C. Gürkan, and B. Kell, "Ant colony optimization applied to the bike sharing problem." Carnegie Mellon University.

L. Breiman, "Random forests," *Machine Learning*, vol. 45, pp. 5–32, 2001.

C. Contardo, C. Morency, and L.-M. Rousseau, "Balancing a dynamic public bike-sharing system," *Tech. Rep. CIRRELT-2012-09, CIRRELT*, vol. 4, Montreal, 2012.

D. W. Daddio, and N. Mcdonald. "Maximizing bicycle sharing: an empirical analysis of capital bikeshare usage." University of North Carolina at Chapel Hill, 2012.

M. Dell'Amico, E. Hadjicostantinou, M. Iori, and S. Novellani, "The bike sharing rebalancing problem: mathematical formulations and benchmark instances," *Omega*, vol. 45, pp. 7–19, 2014.

A. Demiriz, K. P. Bennett, and M. J. Embrechts, "Semi-supervised clustering using genetic algorithms," *Artificial Neural Networks in Engineering (ANNIE-99)*, 1999, pp. 809–814.

G. M. Dias, B. Bellalta, and S. Oechsner, "Predicting occupancy trends in Barcelona's bicycle service stations using open data," arXiv preprint: arXiv:1505.03662, 2015, pp. 439–445.

J. Du, R. He, and Z. Zhechev, "Forecasting bike rental demand," Stanford University, 2014.

C. F. Eick, N. Zeidat, and Z. Zhao, "Supervised clustering-algorithms and benefits," in *16th IEEE International Conference on Tools with Artificial Intelligence (ICTAI 2004)*, 2004, pp. 774–776.

G. Erdoğan, M. Battarra, and R. Wolfler Calvo, "An exact algorithm for the static rebalancing problem arising in bicycle sharing systems," *European Journal of Operational Research*, vol. 245, pp. 667–679, 2015.

G. Forestier, P. Gançarski, and C. Wemmert, "Collaborative clustering with background knowledge," *Data & Knowledge Engineering*, vol. 69, pp. 211–228, 2010.

C. Fricker and N. Gast, "Incentives and redistribution in homogeneous bike-sharing systems with stations of finite capacity," *EURO Journal on Transportation and Logistics*, vol. 5, pp. 261–291, 2016.

J. H. Friedman, "Greedy function approximation: a gradient boosting machine," *Annals of Statistics*, pp. 1189–1232, 2001.

J. Froehlich, J. Neumann, and N. Oliver, "Sensing and predicting the pulse of the city through shared bicycling," In IJCAI, vol. 9, pp. 1420–1426. 2009.

D. Gale and L. S. Shapley, "College admissions and the stability of marriage," *The American Mathematical Monthly*, vol. 69, pp. 9–15, 1962.

C. Gallop, C. Tse, and J. Zhao, "A seasonal autoregressive model of Vancouver bicycle traffic using weather variables," *i-Manager's Journal on Civil Engineering*, vol. 1, p. 9, 2011.

Y. Ganjisaffar, R. Caruana, and C. V. Lopes, "Bagging gradient-boosted trees for high precision, low variance ranking models," In Proceedings of the 34th international ACM SIGIR conference on Research and development in Information Retrieval, pp. 85–94. *ACM*, 2011.

P. Geladi and B. R. Kowalski, "Partial least-squares regression: a tutorial," *Analytica Chimica Acta*, vol. 185, pp. 1–17, 1986.

S. Ghosh, P. Varakantham, Y. Adulyasak, and P. Jaillet, "Dynamic repositioning to reduce lost demand in bike sharing systems," *Journal of Artificial Intelligence Research*, vol. 58, pp. 387–430, 2017.

R. Giot and R. Cherrier, "Predicting bikeshare system usage up to one day ahead," In Computational intelligence in vehicles and transportation systems (CIVTS), 2014 IEEE symposium on, pp. 22–29. *IEEE*, 2014.

S. Goodyear, *Mapping the Imbalances of New York's Popular, Troubled Bike-Share*, 2014. Available: www.citylab.com/commute/2014/03/mapping-imbalances-new-yorks-popular-troubled-bike-share/8699/

H. Hernández-Pérez and J.-J. Salazar-González, "A branch-and-cut algorithm for a traveling salesman problem with pickup and delivery," *Discrete Applied Mathematics*, vol. 145, pp. 126–139, 2004a.

H. Hernández-Pérez and J.-J. Salazar-González, "Heuristics for the one-commodity pickup-and-delivery traveling salesman problem," *Transportation Science*, vol. 38, pp. 245–255, 2004b.

S. C. Ho and W. Y. Szeto, "Solving a static repositioning problem in bike-sharing systems using iterated tabu search," *Transportation Research Part E: Logistics and Transportation Review*, vol. 69, pp. 180–198, 2014.

A. Höskuldsson, "PLS regression methods," *Journal of Chemometrics*, vol. 2, pp. 211–228, 1988.

A. Kaltenbrunner, R. Meza, J. Grivolla, J. Codina, and R. Banchs, "Urban cycles and mobility patterns: exploring and predicting trends in a bicycle-based public transport system," *Pervasive and Mobile Computing*, vol. 6, pp. 455–466, 2010.

D. Marcu, "A Bayesian model for supervised clustering with the Dirichlet process prior," *Journal of Machine Learning Research*, vol. 6, pp. 1551–1577, 2005.

S. P. Mohanty, U. Choppali, and E. Kougianos, "Everything you wanted to know about smart cities: the Internet of Things is the backbone," *IEEE Consumer Electronics Magazine*, vol. 5, pp. 60–70, 2016.

M. Rainer-Harbach, P. Papazek, B. Hu, and G. R. Raidl, "Balancing bicycle sharing systems: a variable neighborhood search approach," in *European Conference on Evolutionary Computation in Combinatorial Optimization*, 2013, pp. 121–132.

T. Raviv, M. Tzur, and I. A. Forma, "Static repositioning in a bike-sharing system: models and solution approaches," *EURO Journal on Transportation and Logistics*, vol. 2, pp. 187–229, 2013.

R. Rixey, "Station-level forecasting of bikesharing ridership: station network effects in three US systems," *Transportation Research Record: Journal of the Transportation Research Board*, 2387, pp. 46–55, 2013.

C. Rudloff and B. Lackner, "Modeling demand for bicycle sharing systems–neighboring stations as a source for demand and a reason for structural breaks," Transportation Research Record 2430.1 (2014): 1–11.

J. Schuijbroek, R. Hampshire, and W.-J. van Hoeve, "Inventory rebalancing and vehicle routing in bike sharing systems," 2013.

Y. Şenbabaoğlu, G. Michailidis, and J. Z. Li, "Critical limitations of consensus clustering in class discovery," *Scientific Reports*, vol. 4, p. 6207, 2014.

J. Shu, M. C. Chou, Q. Liu, C.-P. Teo, and I.-L. Wang, "Models for effective deployment and redistribution of bicycles within public bicycle-sharing systems," *Operations Research*, vol. 61, pp. 1346–1359, 2013.

J. Sinkkonen, S. Kaski, and J. Nikkilä, "Discriminative clustering: optimal contingency tables by learning metrics," in *European Conference on Machine Learning*, 2002, pp. 418–430.

P. Vogel, T. Greiser, and D. C. Mattfeld, "Understanding bike-sharing systems using data mining: Exploring activity patterns," *Procedia-Social and Behavioral Sciences*, vol. 20, pp. 514–523, 2011.

X. Wang, G. Lindsey, J. E. Schoner, and A. Harrison, "Modeling bike share station activity: effects of nearby businesses and jobs on trips to and from stations," *Journal of Urban Planning and Development*, vol. 142, p. 04015001, 2015.

H. Wold, "Soft modelling: the basic design and some extensions," *Systems Under Indirect Observation, Part II*, pp. 36–37, 1982.

S. Wold, A. Ruhe, H. Wold, and I. W. Dunn, "The collinearity problem in linear regression. The partial least squares (PLS) approach to generalized inverses," *SIAM Journal on Scientific and Statistical Computing*, vol. 5, pp. 735–743, 1984.

S. Wold, M. Sjöström, and L. Eriksson, "PLS-regression: a basic tool of chemometrics," *Chemometrics and Intelligent Laboratory Systems*, vol. 58, pp. 109–130, 2001.

S. Xiaoyan, Z. Fanggeng, and G. Yancheng, "Genetic algorithm for the one-commodity pickup-and-delivery vehicle routing problem," in *IEEE International Conference on Intelligent Computing and Intelligent Systems*, 2009, pp. 175–179.

H. Xiong, J. Wu, and J. Chen, "K-means clustering versus validation measures: a data-distribution perspective," *IEEE Transactions on Systems, Man, and Cybernetics, Part B (Cybernetics)*, vol. 39, pp. 318–331, 2009.

Y.-C. Yin, C.-S. Lee, and Y.-P. Wong, "Demand prediction of bicycle sharing systems," Stanford University Online. 2012.

J. W. Yoon, F. Pinelli, and F. Calabrese, "Cityride: a predictive bike sharing journey advisor," In Mobile Data Management (MDM), 2012 IEEE 13th International Conference on, pp. 306–311. IEEE, 2012.

6 Indirect Monitoring of Critical Transport Infrastructure

Data Analytics and Signal Processing

Abdollah Malekjafarian, Eugene J. OBrien, and Fatemeh Golpayegani

CONTENTS

6.1 INTRODUCTION

Today's urban transport networks play a key role in supporting economic growth, providing leisure opportunities and promoting social cohesion in cities. Changes in technology, along with urbanization and economic growth, are driving a demand for personal mobility and transportation. It is expected that transport infrastructure systems in future cities will face a challenge to keep pace with these changes. Transport systems will need to adapt and transform their performance.

The vision of smart cities is to achieve safe, environmentally responsible, secure, sustainable, and efficient urban areas including roads, railways, bridges, and tunnels. A city that is capable of monitoring the condition of its critical infrastructure can plan preventive maintenance activities and optimize its resources. Deterioration and damage will inevitably occur in transport infrastructure such as roads and railway systems. Condition monitoring of these structures is necessary for the effective planning of maintenance activities and to minimize unscheduled closures. In most health monitoring techniques for road and railway structures, the common practice is to mount sensors at different positions on the structure. However, the on-site installations are costly, time-consuming, and even dangerous, depending on the location and type of structure. As a result, covering an entire railway or highway network is infeasible at this time. It would need an inordinate number of sensors, power installation or energy harvesting technology for the sensors and an extensive network of data acquisition, processing and communications electronics.

Recently, researchers have shown an increased interest in "indirect" or "drive-by" methods in which the condition or health of the infrastructure is monitored using an instrumented vehicle (Malekjafarian et al. 2015). While still at an early stage of development, this technique has been successfully employed for the monitoring of road pavements (Nasimifar et al. 2016; Carrera et al. 2013), highway and railway bridges (OBrien and Malekjafarian 2016; OBrien et al. 2017) and railway tracks (Quirke et al. 2017). Such an approach is low-cost at the network level and is aimed at reducing the need for any direct installation of monitoring equipment on the structure itself. In these methods, the condition of the infrastructure is assessed based on the time signals measured continuously or periodically in vehicle(s). For example, two years of data measured on an in-service train, is analyzed in (Lederman et al. 2017) for the purpose of railway track monitoring. In another example, data measured on a city bus is used in (Yabe et al. 2014) for the health monitoring of short-and medium-span bridges. A common challenge of indirect monitoring is dealing with large quantities of measured data and how to transform it into helpful information. Continuous measurement from passing vehicles provides a great quantity of data which needs effective methods to analyze it and transform it into meaningful information about the condition of infrastructure.

In this chapter, recent approaches to indirect monitoring of transport infrastructure are introduced and reviewed and recommendations for future development are provided. It also aims to guide researchers in implementing the available drive-by monitoring methods which have good potential. In addition, the use of approaches such as pattern recognition, implemented through machine learning algorithms, are discussed. It is shown how large numbers of time signals obtained from many vehicle passes can be processed and used for condition assessment of structures. The main challenges in data analysis of measured vehicle data for indirect monitoring are discussed and guidance is provided on choosing and implementing the available algorithms and signal processing methods.

6.2 INDIRECT MONITORING OF TRANSPORT INFRASTRUCTURE

Transport infrastructure may consist of highways, railways, bridges, tunnels, etc., which need condition monitoring. Structures and infrastructure deteriorate over time – ground settles under repeated loading, steel is subject to fatigue and oxidation causes materials to deteriorate. Increases in traffic loading, natural disasters, bridge strikes and seismic activity are among the events that may accelerate natural aging processes and reduce service life (Casas et al. 2017). Transport networks are extensive. For example, there are about one million highway bridges and a half million railway bridges in Europe, of which 35% are over 100 years old (Jensen et al. 2014). It is necessary to monitor the condition of these structures to prevent unplanned closures such as the Hammersmith Flyover in the United Kingdom which was closed without notice in 2011 when it failed a structural inspection. An even more tragic and disruptive event was the I-35W Mississippi River Bridge in Minnesota, United States, which collapsed without warning in 2007 (see Figure 6.1).

Visual inspection in the traditional and the most common method of monitoring transport infrastructure. However, it is subjective and therefore highly variable in its effectiveness. It also lacks resolution and in many cases can only detect damage when it is visible. A number of bridge collapses have occurred due to a lack of structural capacity information which suggests that visual inspection alone may not be adequate for structural health monitoring (Chupanit and Phromsorn 2012). This means that there is a necessity for using other complementary methods for the monitoring of transport infrastructure.

Camera systems have also been used for monitoring transport infrastructure. For example, cameras indicate visually if the bridge deck moves following

Figure 6.1 I-35W Mississippi river bridge failure (Willis, 2007).

a strike by a vehicle. Targets are put in the camera's line of vision to allow quantification of the level of movement in the event of a strike. Linear Variable Differential Transformer (LVDT) systems are used for monitoring of deflections in highway and railway bridges. For example, they can be installed at the connections between the vertical supports and the deck. Road owners might install, say, six LVDTs at each support (pier/abutment), to allow two linear measurements in all three directions, allowing for redundancy and robustness. LVDTs have been used over a number of years to measure displacement at bridge deck-pier and deck-abutment interfaces.

Vibration-based structural health monitoring methods infer information on the properties of a structure from its dynamic response. In direct monitoring, the structure is usually instrumented with many sensors. If power is not an issue, vibration data may be continuously monitored using signal processing methods. There are many examples of instrumentation of transport infrastructure (Chan et al. 2006; Hodge et al. 2015; Alavi et al. 2016). In most vibration-based health monitoring systems, large numbers of sensors are installed on the structure to monitor the dynamic properties. These approaches, in which sensors are installed directly on the structure, are referred to here as direct methods and the on-site instrumentation may be costly, time-consuming and even dangerous to install and maintain, depending on the location and type of the structure. The common practice is to mount the sensors (accelerometers, strain gauges, displacement transducers, level sensing stations, anemometers, temperature sensors, etc.) at different positions and connect them to a data acquisition system with AC electrical power. In the context of a large transport network, such instrumentations tend to be expensive and it is hardly feasible at this time to instrument entire networks (e.g., bridges, tunnels, etc.). Providing power to the sensors and data acquisition systems is another challenge for conventional methods. Using energy harvesting sensors and long-life batteries may be a solution for this challenge, but they are still in development at the time of writing.

An example of a transport network is in the southwest coast of British Columbia in Canada, a seismically active area which presents risks to the Civil Engineering structures built in it. The British Columbia Smart Infrastructure Monitoring System (BCSIMS) system, developed in 2009 (Kaya et al. 2015), is a good example of a smart infrastructure monitoring system. The main purpose of the BCSIMS project is to monitor key structures and provide rapid damage assessment following a seismic event. It integrates data from 16 instrumented structures (14 bridges and 2 tunnels) and the Strong Motion Network which consists of 170 Internet accelerometer stations. There are 2,594 bridges in total in the province of British Columbia and only 14 of them are fitted with an SHM system. The numbers in Figure 6.2 indicate the number of bridges located in that area. It would clearly be very expensive – effectively impossible – to instrument all the bridges in this system.

It can be concluded that only a low percentage of the bridges within a highway system can be instrumented using direct methods. This tends to consist mostly of long-span bridges which means that short-and medium-span bridges are usually omitted. This may be more significant when highway pavements or railway tracks need to be monitored. For example, the total length of the highway network in New South Wales in Australia is about 13,000 km (Baltzer 2010).

In recent years, many researchers have presented new methods for transport infrastructure monitoring using the dynamic response measured on a passing vehicle (Yang and Yang 2017; OBrien et al. 2017; Malekjafarian et al. 2017). These methods are called drive-by or indirect methods. At network level, such an approach is low-cost and is aimed at reducing the need for any direct installation of equipment on the structure. For example, in the case of BCSIMS, an instrumented vehicle that passes through the whole network can be used instead of instrumenting every structure and section of pavement. It means

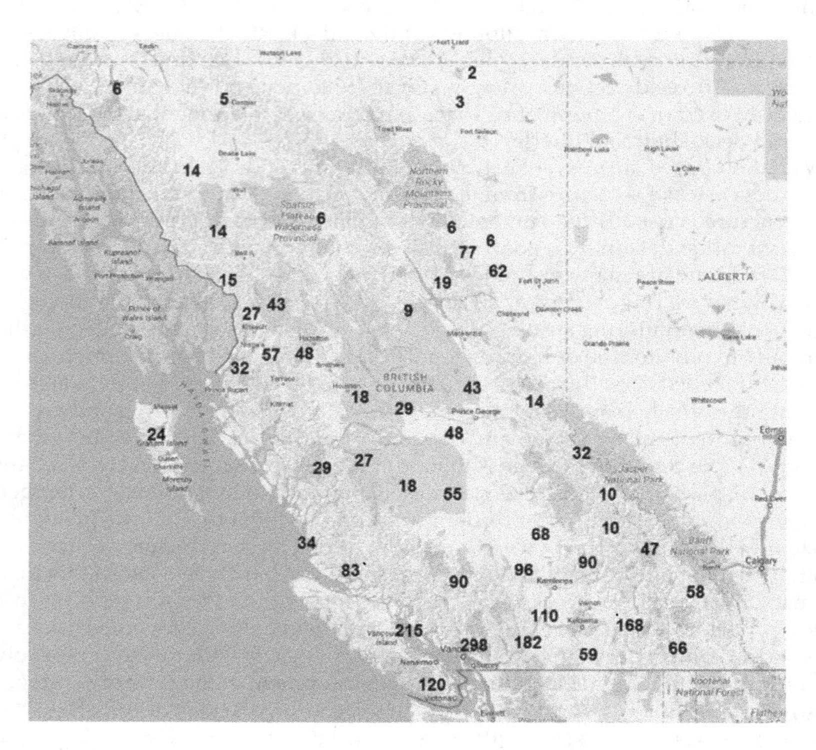

Figure 6.2 The locations of bridges owned by the Ministry of Transportation in British Columbia. The numbers indicate the number of bridges in each location (Kaya et al. 2015).

that all 2,594 bridges could be monitored periodically using one instrumented vehicle. Such an approach could dramatically reduce the cost of monitoring and would clearly increase the effectiveness of the system by considering more than just a small sample of structures.

In this method, the vehicle response is measured continuously as it passes over pavements, bridges and railway tracks. Therefore, a huge quantity of data is measured and needs to be effectively processed to provide reliable information. The data could be stored locally or transferred to a data center using an Internet connection. Processing of the large quantity of data measured on the vehicle to decide whether the structure is healthy or damaged, is a common challenge in indirect methods. Recently, many approaches have been used for processing of indirect measurements (e.g., machine learning and pattern recognition).

6.3 ROAD PAVEMENT APPLICATIONS

Deteriorations in pavements are gradual and steady and typically are addressed by maintenance crew. Increasing traffic volumes and vehicle loads are among the main reasons for pavement deterioration over time (Marecos 2017). There are several defect types for road pavements. Many studies have applied the concept of drive-by measurements to pavement condition monitoring (Flintsch et al. 2012; Baltzer et al. 2010) for different defects.

6.3.1 Foundation Quality

Pavement consists of several layers which may be subject to deterioration over time. Ground Penetrating Radar (GPR) is used for estimating the thickness of layers of road pavement and it can operate at traffic speed (Marecos 2017). It uses an electromagnetic wave at a determined frequency that is propagated vertically through the pavement. The thickness of layers in the pavement is calculated using a part of the wave that is reflected off each layer (Marecos 2017). Many studies have shown that GPR can be used as a reliable and efficient tool for evaluating the integrity of road pavement and to determine whether a pavement section is in a good condition or requires some rehabilitation (Khamzin 2017).

6.3.2 Road Cracking and Surface Roughness

Road pavement distress needs to be detected early to take proper maintenance and rehabilitation actions. Many researchers have used images for road crack detection (Schnebele et al. 2015; Medina et al. 2014; Zalama et al. 2014). When the camera is installed on a passing vehicle traveling at traffic speed, these methods are categorized as drive-by methods. For example, Zalama et al. (2014) use a vehicle instrumented with an imaging system, two inertial profilers, a GPS system and a webcam. The vehicle can travel at a maximum speed of 100 km/h when the imaging system is taking images of the pavement surface. They propose an image classification method for detecting longitudinal and transverse cracks in the road surface. The use of laser measurements is another method of pavement monitoring at high speeds (Tsai 2012). For example, a multi-laser profiler uses several laser measurements (e.g., 48 laser sensors) to measure transversal and longitudinal profiles and can combine them to create a 3D profile.

6.3.3 Pavement Stiffness

Evaluation of the structural capacity of pavement is also important in pavement management. The Falling Weight Deflectometer (FWD) is a testing device which is widely used for the evaluation of structural properties of pavement. However, FWD measurement is a stop-and-go process which limits its use at network level (Donovan and Tutumluer 2009). Continuous deflection devices

(Katicha et al. 2014a) measure deflection at driving speed, avoiding the safety implications of stationary measurements (Katicha et al. 2014b). The two main continuous deflection devices are the Rolling Weight Deflectometer (RWD) and the Traffic Speed Deflectometer (TSD). RWD uses several lasers to measure the deflection under the weight of the vehicle axles. In a TSD, Doppler lasers are used to measure the first derivative of deflection with respect to time, i.e., velocity. The deflections are calculated by post-processing of these measured velocities (Rasmussen 2002). Many studies have been done to infer pavement condition from measured deflections (Donovan and Tutumluer 2009; Malekjafarian et al. 2017). Figure 6.3 shows a sample of data measured by a TSD in Denmark. It shows the data measured in three consecutive years, 2005, 2006 and 2007. Some repair work was carried out between the 2005 and 2006 measurements. The influence of this work on the quality of the pavement can be clearly seen in the figure between 6 and 7.5 km – deflections under the passing load are greatly reduced after the first year, suggesting a stiffer pavement (which results in fewer strain exceedances used in Miner's Rule). It means that a stiffer pavement is more durable and can take greater load without failing. Deflection of pavement under a fixed load is correlated with its stiffness and its bearing capacity. As a result, the pavement deflections measured by these devices is an indicator of bearing capacity.

Road authorities in Australia used a TSD to test an 18,000 km road network in 5 months (Baltzer et al. 2010). It results in a very big quantity of data to be processed. Some parts of the measurements are validated by other pavement inspection methods and devices. The bearing capacity of the pavement is labeled as good, fair and poor using green, yellow and red colors respectively, as shown in Figure 6.4.

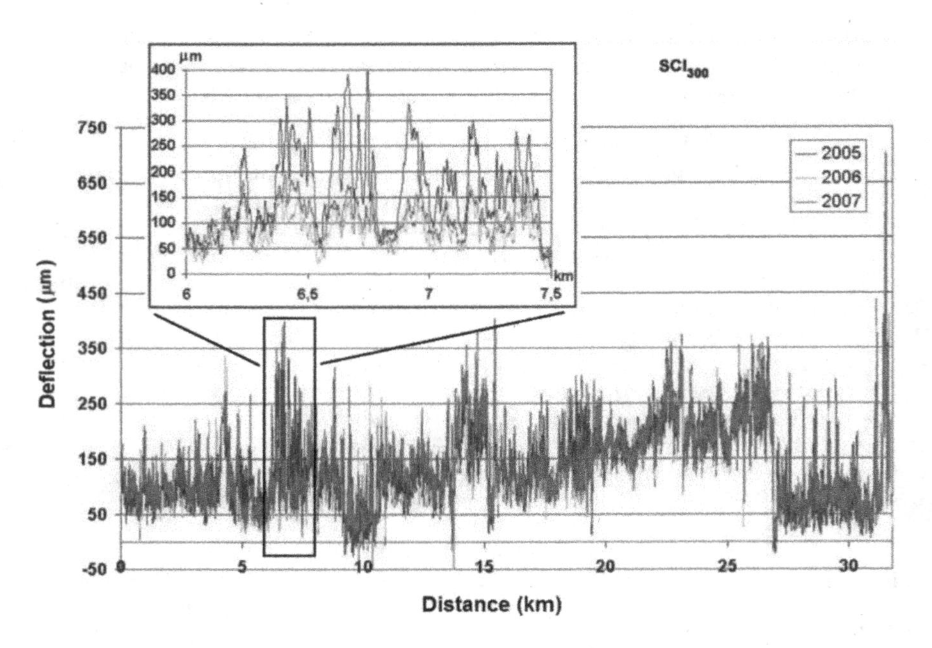

Figure 6.3 Continuous deflection under applied load as an indicator of bearing capacity – a case study in Denmark (Greenwood 2018).

Figure 6.4 Pavement classification using TSD measurements (Greenwood 2018).

6.4 RAILWAY TRACK APPLICATIONS

It is necessary to periodically monitor railway track condition to detect possible faults at an early stage. This ensures the safety of the track and allows maintenance activities to be optimally managed. There are several types of railway track defect; rail cracks, track settlement, landslides onto track, hanging sleepers, etc. Currently, visual inspection ("walking the track") and using Track Recording Vehicles (TRVs) (Figure 6.5) are the most common methods used (Lederman et al. 2017). Considering the total length of railway track in a network, it is labor intensive and expensive to perform frequent (daily in some cases) visual inspections. Further, while a visual system, human or using image analysis, can detect major faults, it may not be reliable for other, more subtle faults. TRVs provide more

Figure 6.5 Track Recording Vehicle in Russia (Anisimov 2006).

comprehensive information as they pass over the whole network and provide useful data on the track such as; longitudinal profile of left and right rail, alignment of left and right rail, track gauge (dual measurement), cross level, twist, curvature and curve radius, gradient and position using GPS. However, they are expensive vehicles and do not always travel at high-speed so they may cause disruption to regular services. This limits the frequency of use of these vehicles which may be as little as twice per annum. To improve this, the frequency of measurement needs to be significantly increased. This can be achieved by replacing the current specialist TRVs with sensing systems on regular passenger or freight trains. These do not disrupt service provision and can provide real-time data on track condition, facilitating more timely responses to developing track failures.

Recently, the use of in-service trains has been proposed for railway track monitoring by several researchers (Lederman et al. 2017; Bocciolone et al. 2007; Molodova et al. 2007; Malekjafarian et al. 2017). In this method, an operational train is instrumented with several sensors (e.g., accelerometer, gyro, GPS, etc.). The sensors are usually installed on the train axle box or bogie and the responses are measured continuously.

Lederman et al. (Lederman et al. 2017) collect an extensive dataset over a 2-year period from an operational train. The train is instrumented using two uni-axial accelerometers inside the cabin, one tri-axial accelerometer on the bogie and a GPS system to track the train location. Responses are measured from hundreds of passes through a sample of 30 km of rail track. The "energy" of the measured signals is plotted in the space domain and a feature detection method is proposed to identify any changes that may indicate track degradation. They investigate a feature extraction method using four features from the raw data. Then they use supervised classification to determine which feature provides the best results. Cumulative sum chart control (CUSUM), generalized likelihood ratio and simplistic Haar filter methods are used for change detection.

In another study, Lederman et al. (Lederman et al. 2017a) propose a novel analysis technique for track monitoring that exploits the scarcity inherent in train-vibration data. They show that there are many bumps in the measured acceleration signals which are related to the track's locations most excite the train. They suggest that the position of the train can be found relative to the sparse "bumps" which helps with the alignment of multiple measurements. They also monitor changes in the size of bumps in consecutive passes to determine if the track is changing.

A data fusion approach is proposed in (Lederman et al. 2017b) for track monitoring from multiple in-service trains (Figure 6.6). This approach employs measurements from multiple sensors in passes of multiple trains. It first minimizes the position offset errors using a data alignment method and then fuses the data using a novel Kalman filter that weights the data according to its estimated reliability. A dataset collected from two instrumented trains operating over a

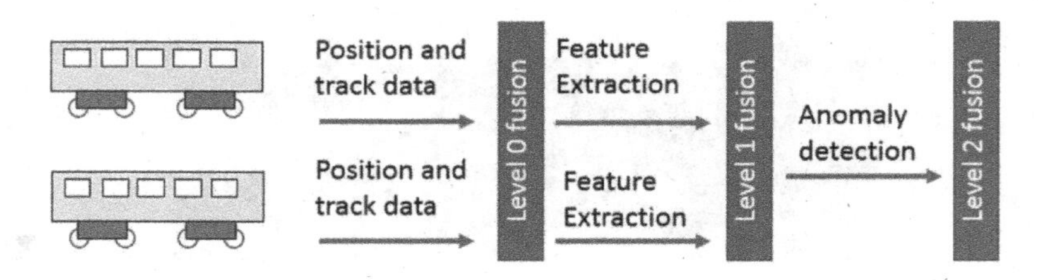

Figure 6.6 The fusion approach proposed by Lederman et al. (2017).

one-year period is used to test the concept. Data from two operational trains passing over a faulty joint, before and after its repair, was used to detect when the repair occurred.

Quirke (2017) employs an inverse technique whereby inertial measurements in an in-service train are used to find the longitudinal track profile that generated the measurement. Testing took place on the Dublin–Belfast line in Ireland. Measurements were recorded for a total of 57 return journeys on the line. The measurements were correlated to particular sections of track using associated GPS data. Bogie vertical acceleration and angular velocity were extracted through an area of known track settlement which was surveyed using traditional techniques during the measuring period. Cross Entropy optimization is used to find the track longitudinal profile that generated the vehicle model responses that best match the measured signals. A good match is achieved between the inferred longitudinal profiles and the surveyed track profile, albeit with an overestimation of the profile elevations.

Issues such as algorithm efficiency, accuracy of vehicle properties, improved representation of non-linear suspension in the vehicle model and knowledge of fuel and passenger loading may all help improve the accuracy and reproducibility of the drive-by railway track monitoring methods. Aside from reproducibility and scaling issues, these methods can be readily implemented using low-cost inertial sensors and simple numerical modeling.

6.5 BRIDGE APPLICATIONS

In recent years, an increasing number of researchers have used indirect measurements to monitor the health condition of bridges (Malekjafarian et al. 2015). Just as in British Columbia in Canada, Japan is vulnerable to seismic activity and rapid monitoring of bridges is an active area of research. Miyamoto and Yabe (2012; 2014) suggest that all bridges should be monitored for early detection of anomalies and appropriate maintenance activates. They propose the use of a city bus to monitor city bridges on a continuous basis. An overview of the concept is shown in Figure 6.7. They instrumented a city bus, which passed regularly over

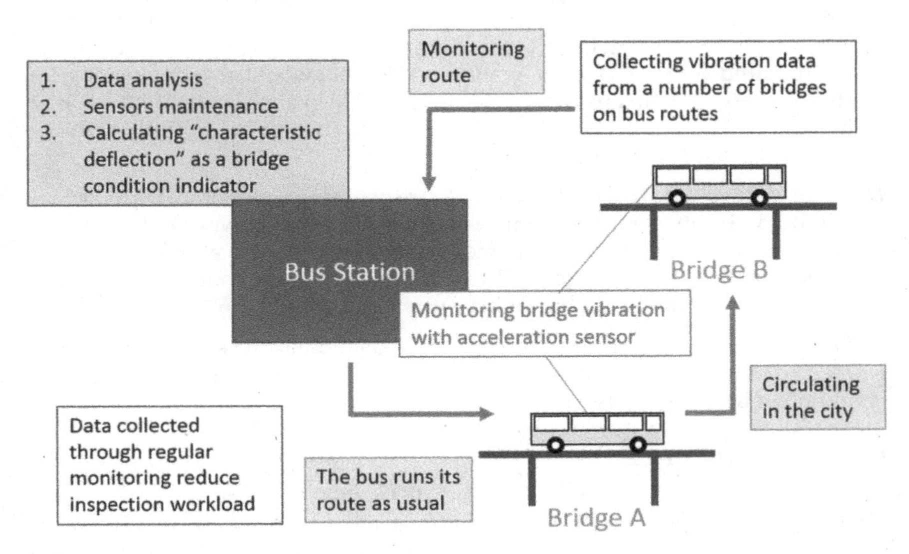

Figure 6.7 Drive-by monitoring system proposed by Miyamoto et al. (2017).

the bridges in the network. The bus passes over a particular bridge at a speed of 30 km/h three times and 40 km/h another three times. It is shown that the response measured at each speed is repeatable. It is suggested that any anomaly in the bridge can be detected by repeating the passes on a regular basis and comparing the acceleration signals measured at the rear axle.

Cerda et al. (2014) present experimental results from a laboratory scale model of vehicle bridge interaction. They run the model vehicle 30 times at eight different velocities and propose a classification system for indirect damage detection. The classification process is a form of supervised learning and consists of a feature extractor and a classifier. The feature extraction is developed in the frequency domain. It takes a feature vector as input and outputs a class label. It is suggested that the Kernel support vector machine classifier can classify healthy and damaged bridges. Chen et al. (2014) study a semi-supervised multi-resolution classification method for indirect bridge health monitoring. They use a framework to deal with non-stationarities in the signal to extract the features in time-frequency regions. Besides the progress that has been made in damage detection methods, big data analytic methods are an increasingly important topic for indirect bridge monitoring purposes.

6.6 VEHICLE MANAGEMENT

Three options are proposed for the vehicles to be used in a drive-by monitoring system for transport infrastructure:

Use a general population of vehicles: A general population can be a few instrumented trucks traveling in a highway network or an entire city network. This would require a system of communications to extract the necessary information (assuming the vehicles are fitted with accelerometers). There would be challenges of permissions, variability in the truck properties, etc. This option is clearly different for railway systems where there are fewer operators (sometimes only one) and scheduled routes.

Use a dedicated fleet of vehicles: For example, a fleet of buses, public service vehicles or in-service trains can be instrumented. As these vehicles are already available in the system, there is less concern about securing permission from the required vehicles. A good quantity of the data would also be available for this strategy as there are many vehicles available in such fleets. With some exceptions, railway fleets tend to be more centrally owned/controlled which makes it easier to employ them.

Use a specialist vehicle: The TSD or the TRV are good examples of this option. Although the instrumented vehicle might be very expensive, it provides more accurate data, such as laser measurements. There are already specialist vehicles in both highway and railway networks in many countries. The challenge then is to integrate them into a system.

Each of these options can be used in a drive-by monitoring system. In some cases, a combination of them may provide the best information regarding the condition of transport infrastructure.

6.7 DATA ANALYTICS

6.7.1 Internet of Things

The Internet of Things (IoT) is fast becoming a key instrument for monitoring a vast array of diverse systems due to its potential to be integrated into smart systems. The "Internet" is generally known as the global system that uses the TCP/

IP protocol suite to interconnect different computer networks (Tokognon 2017). "Things" in this context refers to objects capable of sensing and collecting data. The IoT can therefore be defined as a global system in which objects equipped with sensors are integrated into the information network using intelligent interfaces (Tokognon 2017). In recent years, there has been an increasing interest in using IoT for SHM of different structures (Zhang et al. 2013; Myers et al. 2016).

Figure 6.8 shows a conceptual SHM framework based on the IoT defined by Tokognon et al. (2017). In this system, IoT technologies are used to integrate the system with the Internet to create interaction between smart sensors and data management platforms. An IoT-based system can perform the entire process of structural monitoring, modeling and management. It can also provide real-time information and early warning and facilitate more effective decision making.

This technology can be implemented in the context of indirect monitoring of transport infrastructure. Figure 6.9 shows the proposed IoT approach for a transport system. In this framework, the vehicles (trucks and trains) are instrumented with smart sensors. The sensors are interacting with a data management platform through the Internet connection. As a result, by analyzing the measured data, a health map of the entire network is created and kept up to date using periodic measurements.

6.7.2 Data Analytics Challenges

Data science is fast becoming a key instrument in reshaping the traditional data-based engineering fields. Data analytics methods have the potential to unify experiment, theory and computation in any data-based field including structural health monitoring of transport infrastructure. Cao (2017) shows how a variety of goals and approaches can be used in a data-to-decision process. Figure 6.10 shows three data stages including past data, present data and future data, as the main components of a decision-making process.

In the first stage, a historical analytics approach is used to explore "what happened" in the data and to provide an insight into "how and why it happened" using "Modeling and experimental design." For example, it can contain past

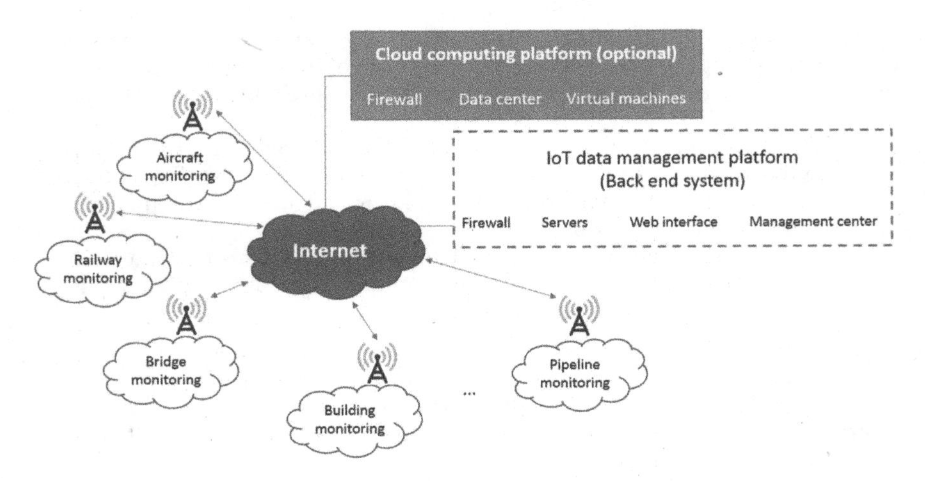

Figure 6.8 The SHM framework based on IoT defined by Tokognon et al. (2017).

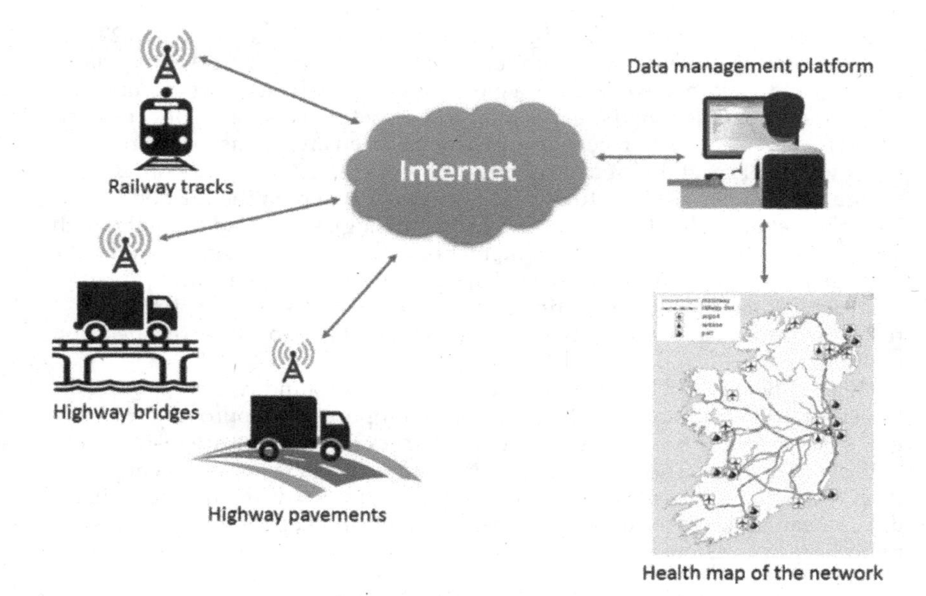

Figure 6.9 The concept of the drive-by monitoring of transport infrastructure.

condition of transport infrastructure based on predefined indicators. The second stage (present data) includes alerts generated in response to particular patterns related to suspicious events in the data, which provides further critical insights into decision-making and real-time risk management processes. For example, the deflection measured by a TSD from some parts of a highway may show a large difference compared to its past data which generates an alert. In the last stage, predictive analytics is employed to predict "what will happen" in the future. The main goal of this stage is estimating the occurrence of future events, and helping to take a right decision for early prevention. For example, if the processed measurements of a TSD exceed a certain threshold, a repair activity needs to be planned. These three data analytics stages enable decision-makers to make an appropriate decision that takes account of the current condition (healthy or damaged, damage levels, etc.) of the structure and take the best next action.

A conceptual framework adapted from Cao (2017), is proposed as a data analytics approach to indirect monitoring of transport infrastructure (Figure 6.11). It summarizes the main challenges in the field from a data science point of view, addressing big data complexities.

The main challenge is establishing mathematical and statistical foundations, to explore the existing theoretical foundation and identify and describe such complexities for a successful data-driven discovery.

- Behavioral and event processing is concerned about modeling and processing of the behaviors and events in a system. In transport systems, structural behavior of structures under healthy and damaged conditions are considered. These behaviors can be modeled using static and dynamic reactions of structures to operational loading in these conditions.
- Data storage and management issues need to be addressed in any data-driven system. Finding a suitable data storage and management process is

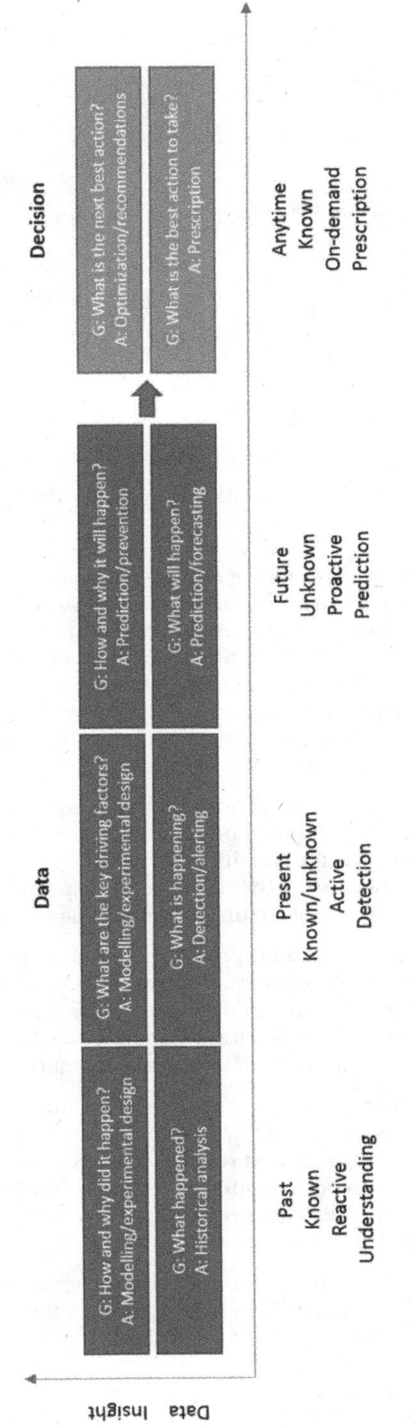

Figure 6.10 Data to insight to decision process (G: analytics goal, A: approach) (Cao 2017).

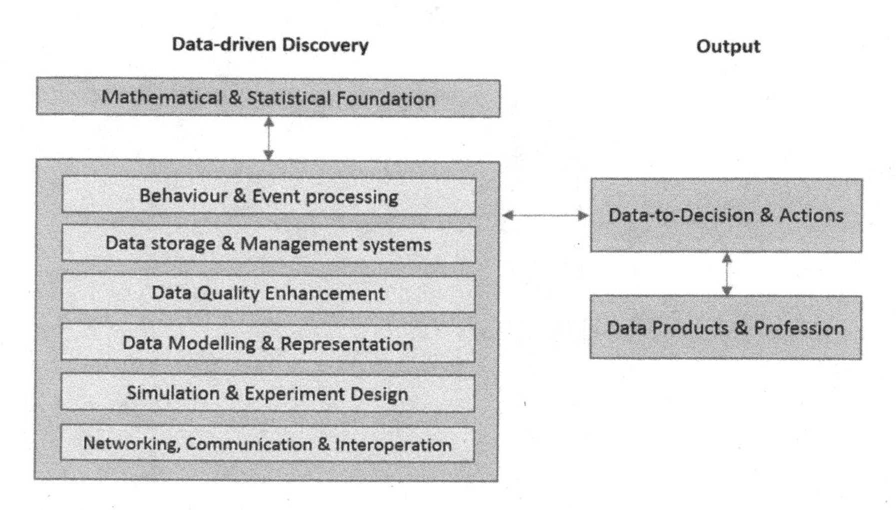

Figure 6.11 Data science challenges for indirect monitoring of transport infrastructure.

also essential in drive-by monitoring of transport systems. The data can be stored in a storage system installed on the passing vehicle and then transferred to the data management unit. As proposed in Figure 6.9, the data can be transferred online through the Internet connection and stored in data management platforms. However, online processing of the data can enhance data storage and management by decreasing the amount of data to be stored. This is achieved by producing and storing only meaningful pieces of data in the form of feature-location, alert-location, etc.

- In most cases, the measured data is not usable as they are stored. It is necessary to use some data cleaning methods to ensure the data quality. For example, some drive-by bridge classification methods only work at a certain vehicle speed range. This means that if the vehicle speed is outside that range, the measured data is not usable and should be excluded.
- Data modeling and representation is concerned with representing data in a comprehensible form. The health map suggested in Figure 6.9, is a good example of addressing this challenge. It contains the map of the considered transport network in which the lines are shown with different colors representing the health condition for each location in the network.
- The optimum experiment situation may depend on many parameters. In some cases, simulations are useful for developing an effective experimental strategy. For example, the measured response of a vehicle passing over a bridge can be estimated using a simulated response from a finite element model. The simulation results may provide useful information such as the optimum vehicle speed, temperature effect, etc.

Networking and communication establishes the structure for distributing the data analytics workload across different hubs/teams. In the context of this chapter, this can be mapped to the distribution of vehicles on the transport network. By implementing IoT into drive-by monitoring systems, and establishing proper communication infrastructure, vehicle-to-vehicle communication and transmission of data to data management platforms is feasible. The data management

Table 6.1: Summary of the State of the Art in Infrastructure Monitoring

Road Pavements	Railway Track	Bridges (Road and Rail)
Ground penetrating radar used to detect foundation issues	Visual inspection (walking the track) still routinely used	Instrumentation of individual bridges is common in large structures but generally only used for research studies of smaller bridges
Cameras or profilometers with lasers used to detect cracking or increased surface roughness	Track Recording Vehicles widely used for occasional detailed inspections	Indirect monitoring is a relatively new concept but has great potential
Traffic Speed Deflectometer or Rolling Wheel Deflectometer used to detect areas with reduced stiffness	Instrumentation in regular trains is a promising approach	

platform can provide some critical information for the vehicle based on its location. For example, the vehicle can be asked to change its route and perform more measurements from particular components of the transport infrastructure. This decreases the cost of monitoring and improves the quality of the measured data.

6.8 CONCLUSION

In this chapter, the most recent big data applications for drive-by monitoring of transport infrastructure are reviewed. The applications in three main categories of transport infrastructure are discussed; highway pavements, railway tracks and bridges. The state of the art in this field is summarized in Table 6.1.

The advantages of drive-by methods over direct methods are addressed for these applications. It is suggested that a group of vehicles, a fleet of vehicles or a specialized vehicle can be alternative strategies. Based on the local policies, procedures and possibilities, one of these options or a combination of them can be employed for drive-by monitoring purposes. A conceptual framework is proposed for indirect monitoring of transport infrastructure. In this framework, smart sensors are installed in vehicles. The sensors are connected to a data management platform using an Internet connection. The condition of the entire network is monitored and kept up to date by continuous processing of the measured data. The importance of integrating data analytics methods for an effective drive-by transport infrastructure platform is clarified and the main challenges are discussed. The potential for the implementation of IoT concepts into this application is highlighted as a basis for the future of structural health monitoring of transport infrastructure. IoT provides effective communications infrastructure which enables vehicle-to-vehicle communication and transmission of measured data to data a management platform.

ACKNOWLEDGMENT

The authors wish to acknowledge the financial support received from Science Foundation Ireland under the US–Ireland Partnership Scheme.

REFERENCES

A.H. Alavi, H. Hasni, N. Lajnef, K. Chatti, Continuous health monitoring of pavement systems using smart sensing technology, *Construction and Building Materials*, 114 (2016) 719–736.

V. Anisimov, 2006, https://commons.wikimedia.org/wiki/File: MTKP.JPG.

S. Baltzer, D. Pratt, J. Weligamage, J. Adamsen, G. Hildebrand, Continuous bearing capacity profile of 18,000 km Australian road network in 5 months, *24th ARRB Conference*, Melbourne, Australia (2010).

M. Bocciolone, A. Caprioli, A. Cigada, A. Collina, A measurement system for quick rail inspection and effective track maintenance strategy, *Mechanical Systems and Signal Processing*, 21 (2007) 1242–1254.

F. Carrera, S. Guerin, J.B. Thorp, By the people, for the people: the crowdsourcing of "streetbump," an automatic pothole mapping app, *International Archives of the Photogrammetry*, 40-4-W1 (2013) 19–23.

J.R. Casas, J.J. Moughty, Bridge damage detection based on vibration data: Past and new developments, *Frontiers in Built Environment*, 3 (2017) 4.

F. Cerda, S.H. Chen, J. Bielak, J.H. Garrett, P. Rizzo, J. Kovacevic, Indirect structural health monitoring of a simplified laboratory-scale bridge model, *Smart Structures and Systems*, 13 (2014) 849–868.

T.H.T. Chan, L. Yu, H.Y. Tam, Y.Q. Ni, W.H. Chung, L.K. Cheng, Fiber Bragg grating sensors for structural health monitoring of Tsing Ma bridge: Background and experimental observation, *Engineering Structures*, 28 (2006) 648–659.

S.H. Chen, F. Cerda, P. Rizzo, J. Bielak, J.H. Garrett, J. Kovacevic, Semi-supervised multiresolution classification using adaptive graph filtering with application to indirect bridge structural health monitoring, *IEEE Transactions on Signal Processing*, 62 (2014) 2879–2893.

P. Chupanit, C. Phromsorn, The importance of bridge health monitoring, *International Science Index*, 6 (2012) 135–138.

Greenwood Engineering (2018), https://www.greenwood.dk/tsd.php.

P. Donovan, E. Tutumluer, Falling weight deflectometer testing to determine relative damage in asphalt pavement unbound aggregate layers, *Journal of the Transportation Research Board* (2009), 2104, 12–23.

G.W. Flintsch, B. Ferne, B. Diefenderfer, S. Katicha, J. Bryce, S. Nell, Evaluation of traffic-speed deflectometers, *Transportation Research Record* (2012), 2304, 37–46.

V.J. Hodge, S. O'Keefe, M. Weeks, A. Moulds, Wireless sensor networks for condition monitoring in the railway industry: A survey, *IEEE Intelligent Transportation Systems*, 16 (2015) 1088–1106.

J.S. Jensen, M. Sloth, P. Linneberg, B. Paulsson, L. Elfgren, MAINLINE: MAINtenance, renewaL and Improvement of rail transport iNfrastructure to reduce Economic and environmental impact, *International Conference of Bridge Maintenance, Safety and Management*: 07/07/2014-11/07/2014, CRC Press, Taylor & Francis Group (2014) 1056–1063.

S.W. Katicha, J. Bryce, G. Flintsch, B. Ferne, Estimating "true" variability of traffic speed deflectometer deflection slope measurements, *Journal of Transportation Engineering*, 141 (2014a).

S.W. Katicha, G. Flintsch, J. Bryce, B. Ferne, Wavelet denoising of TSD deflection slope measurements for improved pavement structural evaluation, *Computer-Aided Civil and Infrastructure Engineering*, 29 (2014b) 399–415.

Y. Kaya, C. Ventura, S. Huffman, and M. Turek. (2017). British Columbia smart infrastructure monitoring system. *Canadian Journal of Civil Engineering*, 44(8), 579–588.

A.K. Khamzin, A.V. Varnavina, E.V. Torgashov, N.L. Anderson, L.H. Sneed, Utilization of air-launched ground penetrating radar (GPR) for pavement condition assessment, *Construction and Building Materials*, 141 (2017) 130–139.

G. Lederman, S.H. Chen, J. Garrett, J. Kovacevic, H.Y. Noh, J. Bielak, Track-monitoring from the dynamic response of an operational train, *Mechanical Systems and Signal Processing*, 87 (2017) 1–16.

G. Lederman, S.H. Chen, J.H. Garrett, J. Kovacevic, H.Y. Noh, J. Bielak, A data fusion approach for track monitoring from multiple in-service trains, *Mechanical Systems and Signal Processing*, 95 (2017a) 363–379.

G. Lederman, S.H. Chen, J.H. Garrett, J. Kovacevic, H.Y. Noh, J. Bielak, Track monitoring from the dynamic response of a passing train: A sparse approach, *Mechanical Systems and Signal Processing*, 90 (2017b) 141–153.

A. Malekjafarian, D. Martinez, E.J. OBrien, Pavement Condition Measurement at High Velocity using a TSD, *27th Annual European Safety and Reliability Conference (ESREL 2017)*, Portoroz, Slovenia (2017).

A. Malekjafarian, P.J. McGetrick, E.J. OBrien, A review of indirect bridge monitoring using passing vehicles, *Shock and Vibration* (2015).

A. Malekjafarian, E.J. OBrien, D. Cantero, Railway track monitoring using drive-by measurements, *The 15th East Asia-Pacific Conference on Structural Engineering and Construction (EASEC-15)*, Xi'an, China (2017).

A. Malekjafarian, E.J. OBrien, On the use of a passing vehicle for the estimation of bridge mode shapes, *Journal of Sound and Vibration*, 397 (2017) 77–91.

V. Marecos, S. Fontul, M. de Lurdes Antunes, M. Solla, Evaluation of a highway pavement using non-destructive tests: Falling weight deflectometer and ground penetrating radar, *Construction and Building Materials* (2017), Vol. 154.

R. Medina, J. Llamas, E. Zalama, J. Gomez-Garcia-Bermejo, Enhanced automatic detection of road surface cracks by combining 2D/3D image processing techniques, *IEEE International Conference on Image Processing* (2014) 778–782.

A. Miyamoto, J. Puttonen, A. Yabe, Long term application of a vehicle-based health monitoring system to short and medium span bridges and damage detection sensitivity, *Engineering*, (2017), 9, 68.

A. Miyamoto, A. Yabe, Development of practical health monitoring system for short-and medium-span bridges based on vibration responses of city bus, *Journal of Civil Structural Health Monitoring*, Paris, France, (2012), 2, 47–63.

M. Molodova, Z.L. Li, R. Dollevoet, Axle box acceleration: Measurement and simulation for detection of short track defects, *Wear*, 271 (2011) 349–356.

A. Myers, M.A. Mahmud, A. Abdelgawad, K. Yelamarthi, Toward integrating structural health monitoring with Internet of Things (IoT), *IEEE International Conference on Electro Information Technology* (2016) 438–441.

M. Nasimifar, S. Thyagarajan, R.V. Siddharthan, N. Sivaneswaran, Robust deflection indices from traffic-speed deflectometer measurements to predict critical pavement responses for network-level pavement management system application, *Journal of Transportation Engineering*, Grand Forks, ND, (2016) 142.

E.J. OBrien, A. Malekjafarian, A. Gonzalez, Application of empirical mode decomposition to drive-by bridge damage detection, *European Journal of Mechanics A-Solid*, 61 (2017) 151–163.

E.J. OBrien, D. Martinez, A. Malekjafarian, E. Sevillano, Damage detection using curvatures obtained from vehicle measurements, *Journal of Civil Structural Health Monitoring*, 7 (2017) 333–341.

P. Quirke, Drive-by detection of railway track longitudinal profile, stiffness and bridge damage, School of Civil Engineering, University College Dublin, Dublin, Ireland (2017).

P. Quirke, D. Cantero, E.J. OBrien, C. Bowe, Drive-by detection of railway track stiffness variation using in-service vehicles, *Proceedings of the Institution of Mechanical Engineers Part F-Journal of Rail and Rapid Transit*, 231 (2017) 498–514.

S. Rasmussen, J.A. Krarup, G. Hildebrand, Non-contact deflection measurement at high speed, in: E. Tutumluer, I.L. Al-Qadi (Eds.) *The 6th International Conference on the Bearing Capacity of Roads, Railways and Airfields*, CRC Press, Taylor & Francis Group, Lisbon, Portugal (2002) 8.

E. Schnebele, B.F. Tanyu, G. Cervone, N. Waters, Review of remote sensing methodologies for pavement management and assessment, *European Transport Research Review*, 7 (2015).

C.J.A. Tokognon, B. Gao, G.Y. Tian, Y. Yan, Structural health monitoring framework based on Internet of Things: A survey, *IEEE Internet of Things Journal*, 4 (2017) 619–635.

Y.C.J. Tsai, F. Li, Critical assessment of detecting asphalt pavement cracks under different lighting and low intensity contrast conditions using emerging 3D laser technology, *Journal of Transportation Engineering – ASCE*, 138 (2012) 649–656.

M. Willis, 2007, https://commons.wikimedia.org/wiki/File:I35_Bridge_Collapse_4.jpg.

A. Yabe, H. E. Miyamoto, Development of a health monitoring system for short- and medium-span bridges based on public bus vibration, *Bridge Maintenance, Safety, Management and Life Extension* (2014) 207–213.

Y. Yang, J.P. Yang, State-of-the-art review on modal identification and damage detection of bridges by moving test vehicles, *International Journal of Structural Stability and Dynamics*, (2017), 18, 1850025.

E. Zalama, J. Gomez-Garcia-Bermejo, R. Medina, J. Llamas, Road crack detection using visual features extracted by Gabor filters, *Computer-Aided Civil and Infrastructure Engineering*, (2014), 29, 342–358.

H.Y. Zhang, J.Q. Guo, X.B. Xie, R.F. Bie, Y.C. Sun, Environmental effect removal based structural health monitoring in the Internet of Things, *2013 Seventh International Conference on Innovative Mobile and Internet Services in Ubiquitous Computing, IMIS*, Taiching, Taiwan (2013) 512–517.

7 Big Data Exploration to Examine Aggressive Driving Behavior in the Era of Smart Cities

Arash Jahangiri, Sahar Ghanipoor Machiani, and Vahid Balali

CONTENTS

7.1 INTRODUCTION

With the advent of smart cities and recent transportation technology advancements such as connected vehicles (CV), transportation systems are fast growing into cooperative intelligent transportation systems. Such vehicle-to-vehicle (V2V) and vehicle-to-infrastructure (V2I) technologies provide the opportunity to identify vehicle location, motion, and immediate driving context (Kamrani et al. 2017). Applications using CV technology are envisioned to improve four objectives including safety, mobility, operational performance, and environment. Safety applications could potentially address approximately 80% of all police-reported crashes annually (Wassim et al. 2010). Moreover, the fast-growing volume and varying format of collected data through the emerging technologies such as CV, have attracted a strong general interest in big data analytics for smart cities. The name "big data" first emerged in the scientific communities in the mid-1990s. Then it slowly became popular around 2008, and started to be recognized in 2010 (Li et al. 2016). Big data challenges have been identified in storing, managing, processing, analyzing, visualizing, and verifying the quality of data. There is a need to address these challenges in transportation applications to facilitate the safety applications of CV data, and to represent advances in real-field large-scale data analytics applications.

More than half of fatal traffic crashes occur due to aggressive driving according to American Automobile Association Foundation for Traffic Safety (2017). Individual driver's crash risks are mainly attributed to driver background, behavioral factors, and driving style. One of the early studies in drivers' behavior measurements evaluated the qualitative measures such as application of brakes, coasting downhill, use of rearview mirror, failing to signal, and so on as crash risks contributing factors (Weiss 1930). Driving style has been assumed to be related to crash risk, however, it has not been fully explored to establish this relationship. Although there is no consensus regarding aggressive driving definition in the literature (Wang et al. 2015), there is an agreement on the negative effect of aggressive driving style on crash occurrence. In addition to safety applications of understanding driving styles (Sagberg et al 2015), driving style affects the comfort of the passengers, and has a direct impact on the energy consumption (Yu et al. 2012) which corresponds to greenhouse gas (GHG) emissions. This magnifies the importance of driving style classification research.

Driving style reflects individual characteristics such as sensation and risk-taking behavior of the driver. It is affected by certain behaviors such as expediency, aggression, compliance, and the joy of speeding. The driver may engage in some of these behaviors by accident without conscious decision. Technological factors including the vehicle system, alert system of potential hazards, and driving task automation also impact the driving style. One example is intelligent cruise control system that has high influence on longitudinal vehicle control. In general, two methods of self-report and behavior observation/recording have been applied in driving style research (Sagberg et al. 2015) (for examples see Taubman-Ben-Ari et al. 2004; Ishibashi et al. 2007; Knipling et al. 2004; Chen et al. 2013). Over the years, researchers categorized driving styles in various groups. For example, Taubman-Ben-Ari et al. (2004) defines the driving style into eight categories of dissociative, anxious, risky, angry, high-velocity, distress reduction, patient, and careful. One of the recent studies by Sagberg et al. (2015) divided driving styles into three groups of calm, careful, and aggressive, and provided description and associated measures of these driving styles. Among all different driving styles definitions and categorization, aggressive driving style has received most attention in road safety research (Sagberg 2015).

Since the risky driving behaviors have been found to be directly correlated with the rate of crashes (Paleti et al. 2010), a wide range of research studies have explored driving behavior and proposed mechanisms for warnings or alerts drivers for improving safety (Chrysler et al. 2015; Genders and Razavi 2016; Abe and Richardson 2006; Doecke et al. 2015; Goodall et al. 2012; Osman et al. 2015; Sengupta et al. 2007; Yang et al. 2000). Even though these researchers applied driving simulator, algorithm development, and closed course experiments, the usage of real-world big data in modeling driving behavior and style is limited. Recently a few pilot studies collected large-scale real-world data with different sensing technologies. These data can be used to develop models for evaluating drivers' decision-making and detecting extreme driving styles and behaviors. For example, analyzing large-scale behavioral data, the relationships between driving style and factors such as driver (e.g., demographics), trip (e.g., purpose, duration), and vehicle features (e.g., body, fuel, transmission, and powertrain) were investigated (Liu et al. 2015). Another study applied connected vehicle data and proposed a framework for producing warnings for potential hazards (Liu and Khattak 2016); using this framework, extreme behaviors (e.g., hard acceleration) can be identified and warn drivers through the V2V and V2I technology. Using real-field CV data, Ghanipoor Machiani et al. (2017) developed a driver behavior model that predicts risky behavior of drivers at the point of curvature (PC) for different monitoring periods from 2 to 7.5 seconds starting from 8 seconds ahead of PC.

Various driving styles can be explored and evaluated using vehicle kinetic data (i.e., speed, acceleration, yaw). The quantification of the driver style can be performed by monitoring instantaneous driving decisions reflected in the kinetic information. Evaluating the characteristics of aggressive, reckless, or risky driving style, speed has been identified as the main factor in determining a driver's performance (Wali et al. 2017; Kim and Choi 2013). Acceleration has also been used as an intuitive measure to identify aggressive driving. Some researchers have determined certain values of motion-related variables as representation of aggressive driving behavior. For example, in urban driving environments, values of 1.47 m/s^2 and 2.28 m/s^2 were reported as the thresholds for aggressive and extremely aggressive acceleration (Kim and Choi 2015), and De Vlieger et al. (2000) reported acceleration range of 0.85–1.1 m/s^2 being associated with aggressive driving. However, speed is a critical variable that affects the capability of vehicles to accelerate/decelerate. Therefore, when identifying aggressive driving based on acceleration, different speed ranges should be treated separately (X. Wang et al. 2015). An example study using kinetic data was conducted by (X. Wang et al 2015). They used acceleration and vehicular jerk to identify aggressive driving. In their definition, a behavior is considered aggressive if acceleration or vehicular jerk goes beyond one standard deviation across all data points for a certain speed range. In another study by J. Wang et al. (2015) braking process features including maximum deceleration, average deceleration, and kinetic energy reduction were used for risky behavior identification. Depending on the available dataset, some researchers also leveraged information associated with crash and near-crash rate events to classify drivers (J. Wang et al. 2015; Guo and Fang 2013; Soccolich et al. 2011).

Driving style and behavior classification has been performed by statistical and machine learning techniques. For example, Guo and Fang (2013) applied negative binomial regression analysis to assess risk factors associated with individual drivers and K-means clustering method to classify high-risk drivers. Ghanipoor Machiani and Abbas (2016) predicted driver behavior at the onset of yellow signal indication using canonical discriminant analysis. Techniques such as random forest, fuzzy logic inference systems, hidden Markov models, and support vector machines (SVM) have also been identified as promising artificial intelligence algorithms for driving behavior modeling and style analysis (Meiring and Myburgh 2015). Red light running violation behavior modeling for passenger cars and bicyclists was conducted using machine learning methods such as random forest and SVM (Jahangiri et al. 2015; Jahangiri, Rakha 2016; Jahangiri, Elhenawy 2016). Wang and Xi (2016) applied SVM and K-means methodologies to classify drivers into aggressive vs. moderate using a driving simulator data. They applied a supervised learning method in that they labeled participants as aggressive vs. moderate through a questionnaire which was completed before the simulator run.

Tagging a certain driver behavior as aggressive versus non-aggressive is difficult since the collective driving data of an aggressive driver may include only isolated instances of aggressive driving behavior. The variance in driving styles is influenced by several factors such as driving environments and driver psychological factors as well as vehicle's capability. In addition, the aggressive threshold value is different for individuals (W. Wang et al. 2016). This difficulty is even more pronounced when only the kinetic data is available, and no other information such as self-reporting driving questionnaires, driver history, and driver emotional status is available. Therefore, regardless of individual drivers, short trip segments (i.e., driving moments) were used to tag aggressive behavior. The focus of this chapter is on classifying aggressive driving using kinetic information (speed, acceleration, yaw) obtained from a large dataset. A data storing and processing strategy was used to deal with big data in which data were partitioned into several chunks and accessed through a mapping mechanism. Two implementations of an unsupervised learning method, namely K-means, were adopted to develop clustering models and examine the impacts on data processing times when dealing with big data.

The remainder of the chapter is organized as follows. The data used in this study are described in Section 7.2. Subsequently, big data techniques and methods applied are explained in Section 7.3. Section 7.4 presents aggressive driving clustering and spatial distribution results as well as travel time impacts followed by Section 7.5 that provides the conclusions.

7.2 DATA DESCRIPTION

The data used in the study is a part of the safety pilot model deployment (SPMD) which was obtained through the research data exchange (RDE), a data sharing system provided by Federal Highway Administration (Federal Highway Administration 2015). The SPMD was a comprehensive data collection effort which is a part of the Connected Vehicle Safety Pilot Program. This program features real-world implementation of connected vehicle technologies using everyday drivers. The data were collected from approximately 3,000 vehicles equipped with connected vehicle technology traversing the Ann Arbor, MI.

The Safety Pilot environment contains several datasets including DAS1 driving dataset. This dataset includes data collected by a data acquisition system (DAS) developed by University of Michigan Transportation Research Institute instrumented on 66 vehicles (i.e., unique device ID). The DAS collects audio and video data as well as text. However, considering the privacy issues with personally identifiable information in the video and audio data, only text-based data were available through the RDE. The data elements included in this dataset are stored in a series of relational tables each pertaining to a specific set of collected data. The DAS1 dataset was collected with the frequency of 10 Hz and includes four comma-separated files, one of which was used in the present study that contains several variables. A subset of the variables was used in this study as summarized in Table 7.1. For further information on the dataset, we refer the readers to the metadata files provided along with the dataset (USDOT 2014; USDOT 2015).

7.3 METHODOLOGY

The dataset used in this study is relatively large and requires big data techniques to be analyzed. This is especially critical when using a personal computer with typical memory of 4–16 Gb. Even with a 16 Gb of memory, only a portion (50–60% is recommended) should be allocated for data processing as

Table 7.1: Summary of Variables (USDOT 2015)

Name	Description
Device	A unique numeric ID assigned to each DAS
Trip	Count of ignition cycles – each ignition cycle commences when the ignition is in the on position and ends when it is in the off position
Time	Time in centiseconds since DAS started, which (generally) starts when the ignition is in the on position
GpsValidWsu	Communicates whether a GPS data point is valid or not
LatitudeWsu	Latitude from WSU receiver
LongitudeWsu	Longitude from WSU receiver
ValidCanWsu	Valid Vehicle CAN Bus message to WSU
YawRateWsu	Yaw rate from vehicle CAN Bus via WSU
SpeedWsu	Speed from vehicle CAN Bus via WSU
AxWsu	Longitudinal acceleration from vehicle CAN Bus via WSU

the machine would require some memory for running other processes and applications. Cloud computing can also be considered to increase the available memory. While this strategy was not used in this study, it is important to note that the cloud computing technologies have been significantly improving and are now more affordable. In this study, R programming was used for data analysis, model development, and data visualization. R is an open source software environment for statistical computing and graphics that has been identified as one of the best programming languages for data science. Thus, many methods, tools, and platforms have been developing to deal with big data problems in R environment. The following three steps show the steps taken to deal with big data and produce insight.

7.3.1 Data Partitioning

In this study, an approach was adopted in which the large dataset at hand was partitioned into several chunks of data and all chunks were stored in the hard drive. In this type of data storing and processing strategy, objects are created in the R environment, which essentially contain mapping information to the data chunks. This way, out of memory errors do not occur as the data are not loaded into memory and yet any information is accessible through the mapping information that has been created when building the data chunks. It should be noted that this flexibility comes with a cost which is the lower processing speed (Walkowiak 2016). After partitioning the dataset, more than 80 million rows of data were stored in 27 data chunks (known as .ff files) taking about 12.5 Gb of hard disk space. Two variables in the data set indicated if vehicle CAN Bus message and the GPS data was valid. Excluding all invalid data points has reduced the data to about 63 million rows.

7.3.2 Data Extraction

Adopting data partitioning as described in the previous task, allowed us to go through the entire data and split the data into short duration trip segments. Since aggressive driving behavior is the focus of this study, the length of trip segments was assumed to be short (1–10 seconds) as aggressive maneuvers occur in short periods of time. Removing invalid data as described in task 1 resulted in some trips not having continuous trip trajectory data. Hence, all trip segments with incomplete data were excluded which further dropped the number of rows to about 61 million rows. As the data were collected at a high frequency of 10, trip segmentation can significantly reduce the size of data depending on selected trip segment length. For example, using trip segment length of 10 seconds, the number of rows was reduced from 61 million to almost 610,000 rows, with each row representing one trip segment.

As previous studies have indicated (X. Wang et al. 2015), variation of kinetic variables such as acceleration and yaw rate are different for different speed ranges as shown in Figure 7.1. Spread of acceleration and yaw rate are similar within these ranges by visually inspecting this figure. Therefore, cluster analysis was performed separately for three speed ranges of low (10–40 km/h), moderate (40–80 km/h), and high (>80 km/h). Observations with speed of less than 10 km/h were dropped since aggressive behavior is unlikely in this speed range.

7.3.3 Knowledge Discovery

This section adopts cluster analysis to classify short trip segments obtained from task 2 into different categories to identify aggressive driving behavior. Various terms, measures, and definitions have been used to label global driving styles, and there is little consensus on their accurate meaning. However, aggressive

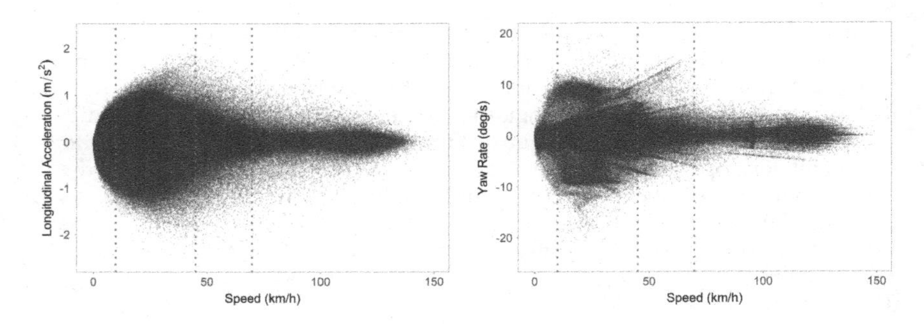

Figure 7.1 Variation of acceleration and yaw for different speed categories.

driving as probably the most researched driving style is considered "typical maladaptive and risk-related behavior in traffic." One may consider distinguishing aggressive driving from normal or non-aggressive driving by classifying all trip segments as either aggressive or normal resulting in two clusters. Similarly, considering more behavior types (e.g., very aggressive; aggressive; normal; conservative) leads to clustering of all trip segments into the respective categories resulting in more than two clusters. Since the trip segments are not labeled as specific driving behavior types, an unsupervised learning approach was used in which number of clusters should be selected to perform the analyses. To determine the best number of clusters, several methods have been used such as the elbow (Ketchen Jr and Shook 1996) and Silhouette (Rousseeuw 1987), and gap statistic (Tibshirani et al. 2001) methods. However, there is no definitive approach and sometimes the number of clusters is selected to meet the desire of the user. At the most general level, driving style can be classified into aggressive/risky and defensive/careful/focused (Sagberg et al. 2015). The present study follows this general distinction to classify driving behavior into aggressive or normal and therefore two clusters were used.

To conduct cluster analysis, this study investigated K-means method as described below.

7.3.3.1 K-Means Method

K-means, firstly introduced in (MacQueen et al. 1967), has been widely used in clustering problems and it has been proven to be simple, efficient, and easy to implement (Jain 2010). Given n observations, the objective function of K-means minimizes within cluster sum-of-squares error using K clusters C_k, $k = 1, \ldots, K$. All observations are assigned to K clusters in such a way so that the objective function is minimized as shown below.

$$\min \sum_{k=1}^{K} \sum_{x_j \in C_k} \left| x_j - \mu_k \right|^2$$

where,

$$\mu_k = \frac{\sum_{x_j \in C_k} x_j}{\left| C_k \right|}$$

Within-group heterogeneity measure, also known as the elbow method, was used with different number of clusters to see the impact of different number of clusters (Lantz 2015). The goal is not to minimize heterogeneity as more clusters usually lead to less heterogeneity, but rather to select a point after which no significant heterogeneity decrease is observed. For implementation, a standard K-means that comes with R statistical software as well as a memory-efficient K-means, called bigKmeans, from the "biganalytics" package that is designed to handle big data were used (Emerson and Jane 2010). The advantage of bigK-means is that it permits a big calculation that might not be feasible computationally with the standard K-means.

7.3.3.2 Scenario Development

Trip segment length variation impacts the number of observations that are used in the unsupervised learning clustering. This study uses different trip segment lengths to investigate the impacts on data processing speed and aggressive driving determination. When choosing smaller lengths, the number of observations can significantly increase. To examine different scenarios, trips lengths of 10, 5, 3, and 1 seconds were investigated. Table 7.2 presents these four scenarios along with their respective number of observations per vehicle speed category. As mentioned earlier, cluster analysis was performed for different speed categories separately.

7.3.3.3 Variable Selection

To identify aggressive driving, variables that are expected to reflect aggressive behavior were selected. In total, six variables were used in model development as presented in Table 7.3. Extreme values of these variables are expected to be associated with aggressive driving. For example, minimum acceleration over a trip segment represents the most extreme value of vehicle deceleration over this trip segment, which could suggest a very hard braking. Maximum (or minimum) yaw rate over a trip segment could suggest a risky vehicle maneuver while abruptly changing a traffic lane or negotiating a horizontal curve at a high speed.

Table 7.2: Scenario Summary

	Trip Segment Length (s)	Number of Observations			
		Low Speed	Moderate Speed	High Speed	Total
Scenario a	10	212,646	100,040	120,650	433,336
Scenario b	5	403,078	207,012	243,283	853,373
Scenario c	3	661,549	347,960	406,598	1,416,107
Scenario d	1	1,973,749	1,050,040	1,222,158	4,245,947

Table 7.3: Variables Used

Variable Name	Description	Variable Name	Description
AxMax	Maximum acceleration	SpeedSD	Speed standard deviation
AxMin	Maximum deceleration	Yaw	Maximum absolute yaw rate
AxSD	Acceleration standard deviation	YawSD	Yaw rate standard deviation

7.4 RESULTS

7.4.1 Cluster Determination

The elbow method was used to see the impact of different number of clusters. As the number of clusters increases within-group sum of squares decreases as presented in Tables 7.4 through 7.7. For almost all scenarios and speed categories, no clear elbow point is recognized. Looking at elbow curves for all cases, if one must select the elbow point, two or three clusters could be realized as the curve slope changes slightly more at these points comparing to the other points. In this study, the number of clusters was chosen to be two as discussed earlier.

Box plots for the six variables (as described in the previous section) are presented in Tables 7.4 through 7.7 that shows variable variation for the two clusters. As an unsupervised learning method was used, it is unknown from the beginning if cluster 1 shows aggressive driving or cluster 2. To investigate this, the mean value of each variable was used to compare the two clusters to realize which cluster contains the most extreme values for that variable (shown in bold). Since the data were not normally distributed the Wilcoxon rank sum (WRS) test was applied as a non-parametric test. The results are summarized in Tables 7.4 through 7.7 for all scenarios and speed categories. Except for one case (variable YawSD in scenario b-low speed category), all p-values were very small that null hypotheses of having equal means were rejected at 95% confidence levels. Considering a scenario with a specific speed category, if mean values of all variables of cluster 1 are more extreme that those of cluster 2, then cluster 1 is identified as aggressive and cluster 2 as normal. However, as can be seen in some cases as presented in Tables 7.4 through 7.7, cluster 1 can contain more extreme values for some variable and less extreme values for other variables. Therefore, in these cases a definitive tagging of clusters as being aggressive or normal may not be appropriate. For example, in scenario b (low speed category in Table 7.5), maximum acceleration, maximum/minimum yaw rate, and yaw rate standard deviation were found more extreme in cluster 1, but minimum acceleration, speed standard deviation, and acceleration standard deviation were higher in cluster 2. This suggests different types of aggressive driving such as abrupt lane change as indicated by high values of yaw rate and hard braking as reflected by low deceleration values, which may not be classified into the same cluster. This is especially the case for shorter trip segments as they are not long enough to include all aggressive driving indicators. For the longest trip segment (10 seconds) a clear distinction of aggressive against normal driving was possible as more extreme values for all variables were in either cluster 1 or 2. Therefore, for this scenario as shown in Table 7.4, cluster 2 was identified as aggressive for low and moderate speed categories and cluster 1 was labeled as aggressive for high speed category.

As presented earlier in Table 7.3, each scenario contained different number of observations. Data processing time of the two K-means method implementations (K-means vs. bigkmeans) was compared to see the impact of number of observations. This comparison is illustrated for each scenario in Tables 7.4 through 7.7. These illustrations show the frequency of processing time for 100 simulations. Also, Figure 7.2 shows how increasing the number of observations impact the processing time for the two implementations. This indicates that the bigkmeans implementation is more efficient especially when the data becomes larger.

Table 7.4: Summary Results for Trip Segments of 10 Seconds

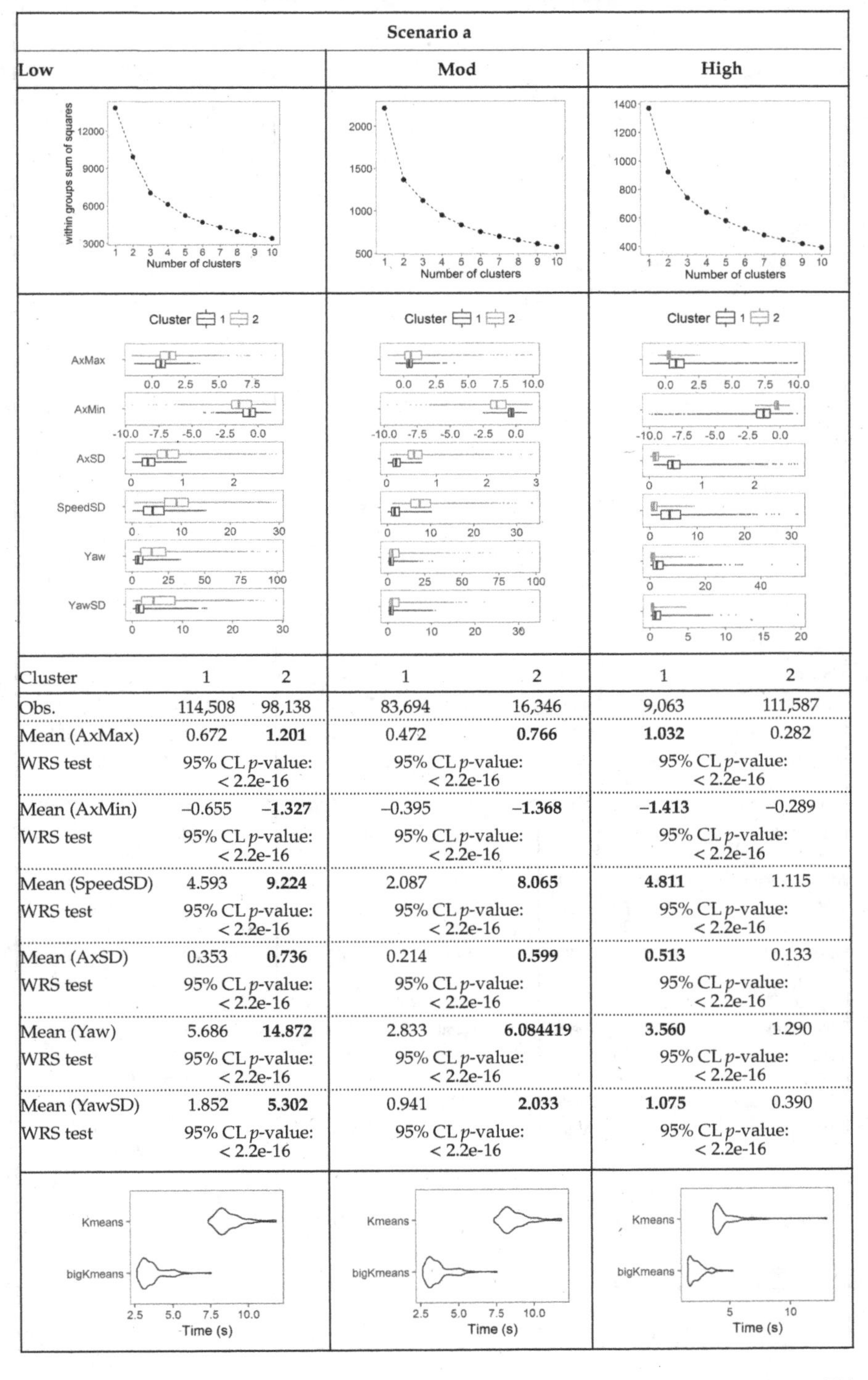

Scenario a						
	Low		Mod		High	
Cluster	1	2	1	2	1	2
Obs.	114,508	98,138	83,694	16,346	9,063	111,587
Mean (AxMax)	0.672	**1.201**	0.472	**0.766**	**1.032**	0.282
WRS test	95% CL *p*-value: < 2.2e-16		95% CL *p*-value: < 2.2e-16		95% CL *p*-value: < 2.2e-16	
Mean (AxMin)	–0.655	**–1.327**	–0.395	**–1.368**	**–1.413**	–0.289
WRS test	95% CL *p*-value: < 2.2e-16		95% CL *p*-value: < 2.2e-16		95% CL *p*-value: < 2.2e-16	
Mean (SpeedSD)	4.593	**9.224**	2.087	**8.065**	**4.811**	1.115
WRS test	95% CL *p*-value: < 2.2e-16		95% CL *p*-value: < 2.2e-16		95% CL *p*-value: < 2.2e-16	
Mean (AxSD)	0.353	**0.736**	0.214	**0.599**	**0.513**	0.133
WRS test	95% CL *p*-value: < 2.2e-16		95% CL *p*-value: < 2.2e-16		95% CL *p*-value: < 2.2e-16	
Mean (Yaw)	5.686	**14.872**	2.833	**6.084419**	**3.560**	1.290
WRS test	95% CL *p*-value: < 2.2e-16		95% CL *p*-value: < 2.2e-16		95% CL *p*-value: < 2.2e-16	
Mean (YawSD)	1.852	**5.302**	0.941	**2.033**	**1.075**	0.390
WRS test	95% CL *p*-value: < 2.2e-16		95% CL *p*-value: < 2.2e-16		95% CL *p*-value: < 2.2e-16	

Table 7.5: Summary Results for Trip Segments of 5 Seconds

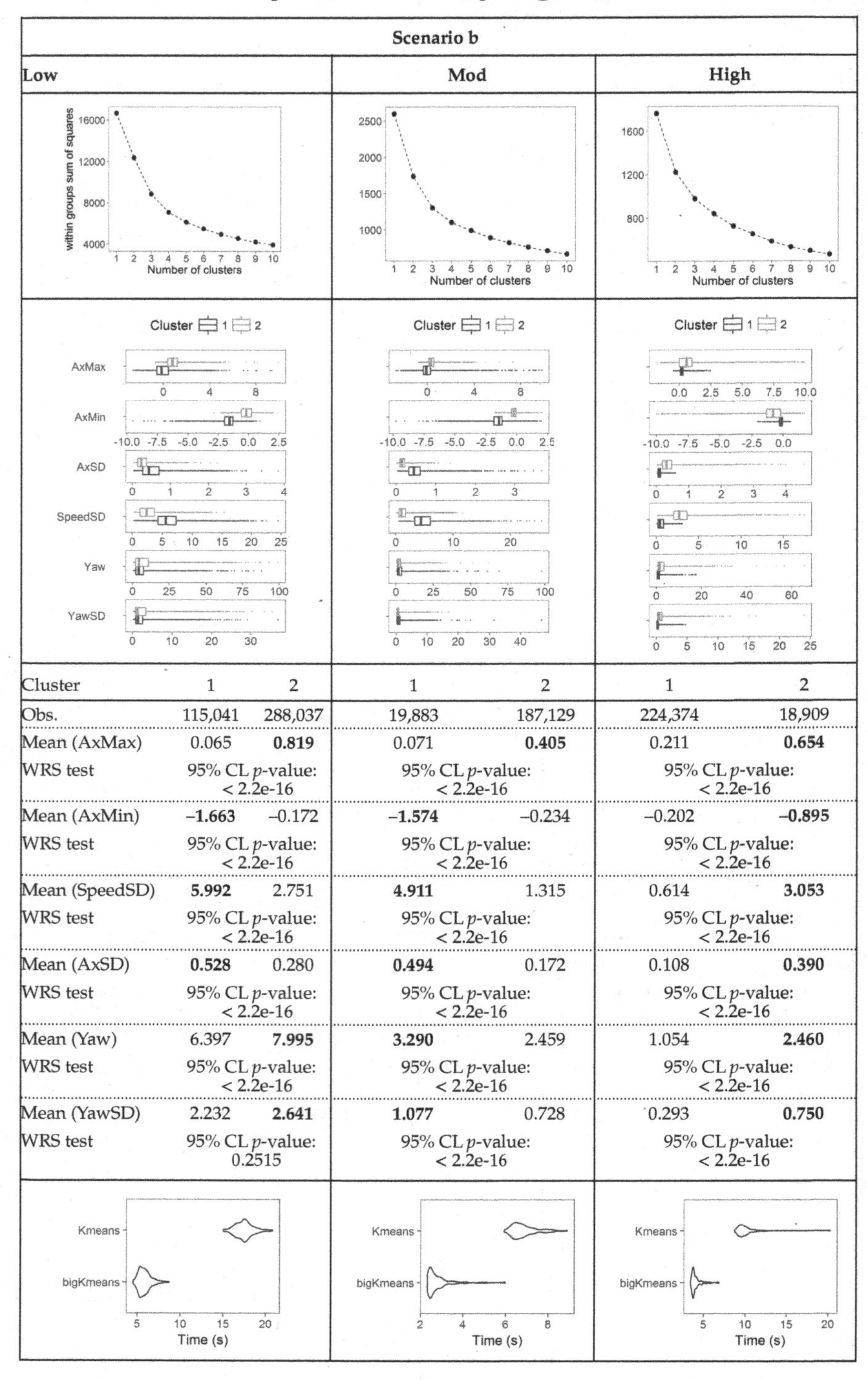

Scenario b						
Low			Mod		High	
Cluster	1	2	1	2	1	2
Obs.	115,041	288,037	19,883	187,129	224,374	18,909
Mean (AxMax)	0.065	**0.819**	0.071	**0.405**	0.211	**0.654**
WRS test	95% CL *p*-value: < 2.2e-16		95% CL *p*-value: < 2.2e-16		95% CL *p*-value: < 2.2e-16	
Mean (AxMin)	**−1.663**	−0.172	**−1.574**	−0.234	−0.202	**−0.895**
WRS test	95% CL *p*-value: < 2.2e-16		95% CL *p*-value: < 2.2e-16		95% CL *p*-value: < 2.2e-16	
Mean (SpeedSD)	**5.992**	2.751	**4.911**	1.315	0.614	**3.053**
WRS test	95% CL *p*-value: < 2.2e-16		95% CL *p*-value: < 2.2e-16		95% CL *p*-value: < 2.2e-16	
Mean (AxSD)	**0.528**	0.280	**0.494**	0.172	0.108	**0.390**
WRS test	95% CL *p*-value: < 2.2e-16		95% CL *p*-value: < 2.2e-16		95% CL *p*-value: < 2.2e-16	
Mean (Yaw)	6.397	**7.995**	**3.290**	2.459	1.054	**2.460**
WRS test	95% CL *p*-value: < 2.2e-16		95% CL *p*-value: < 2.2e-16		95% CL *p*-value: < 2.2e-16	
Mean (YawSD)	2.232	**2.641**	**1.077**	0.728	0.293	**0.750**
WRS test	95% CL *p*-value: 0.2515		95% CL *p*-value: < 2.2e-16		95% CL *p*-value: < 2.2e-16	

Table 7.6: Summary Results for Trip Segments of 3 Seconds

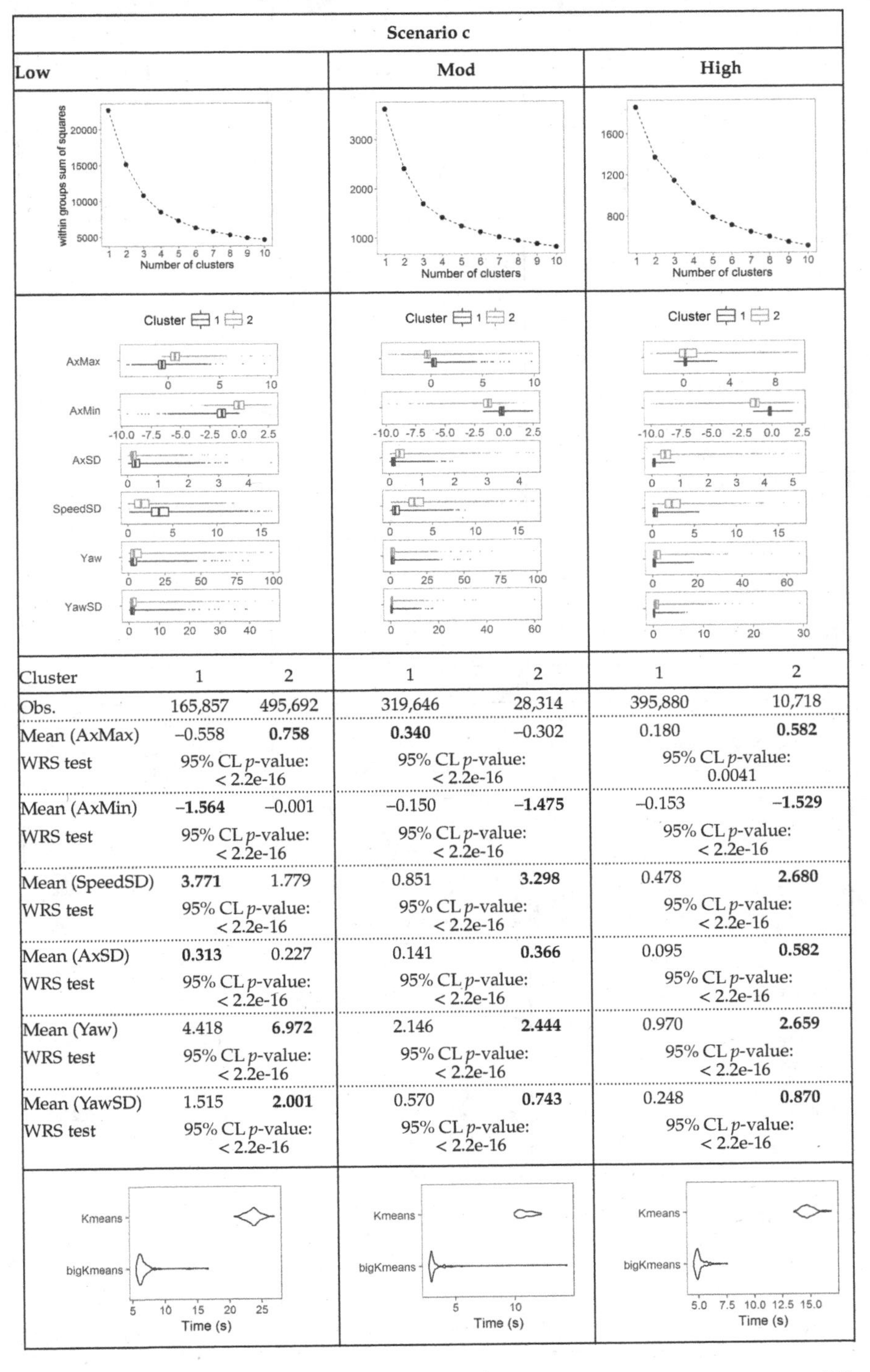

Scenario c						
	Low		Mod		High	
Cluster	1	2	1	2	1	2
Obs.	165,857	495,692	319,646	28,314	395,880	10,718
Mean (AxMax)	–0.558	**0.758**	**0.340**	–0.302	0.180	**0.582**
WRS test	95% CL *p*-value: < 2.2e-16		95% CL *p*-value: < 2.2e-16		95% CL *p*-value: 0.0041	
Mean (AxMin)	**–1.564**	–0.001	–0.150	**–1.475**	–0.153	**–1.529**
WRS test	95% CL *p*-value: < 2.2e-16		95% CL *p*-value: < 2.2e-16		95% CL *p*-value: < 2.2e-16	
Mean (SpeedSD)	**3.771**	1.779	0.851	**3.298**	0.478	**2.680**
WRS test	95% CL *p*-value: < 2.2e-16		95% CL *p*-value: < 2.2e-16		95% CL *p*-value: < 2.2e-16	
Mean (AxSD)	**0.313**	0.227	0.141	**0.366**	0.095	**0.582**
WRS test	95% CL *p*-value: < 2.2e-16		95% CL *p*-value: < 2.2e-16		95% CL *p*-value: < 2.2e-16	
Mean (Yaw)	4.418	**6.972**	2.146	**2.444**	0.970	**2.659**
WRS test	95% CL *p*-value: < 2.2e-16		95% CL *p*-value: < 2.2e-16		95% CL *p*-value: < 2.2e-16	
Mean (YawSD)	1.515	**2.001**	0.570	**0.743**	0.248	**0.870**
WRS test	95% CL *p*-value: < 2.2e-16		95% CL *p*-value: < 2.2e-16		95% CL *p*-value: < 2.2e-16	

Table 7.7: Summary Results for Trip Segments of 1 Second

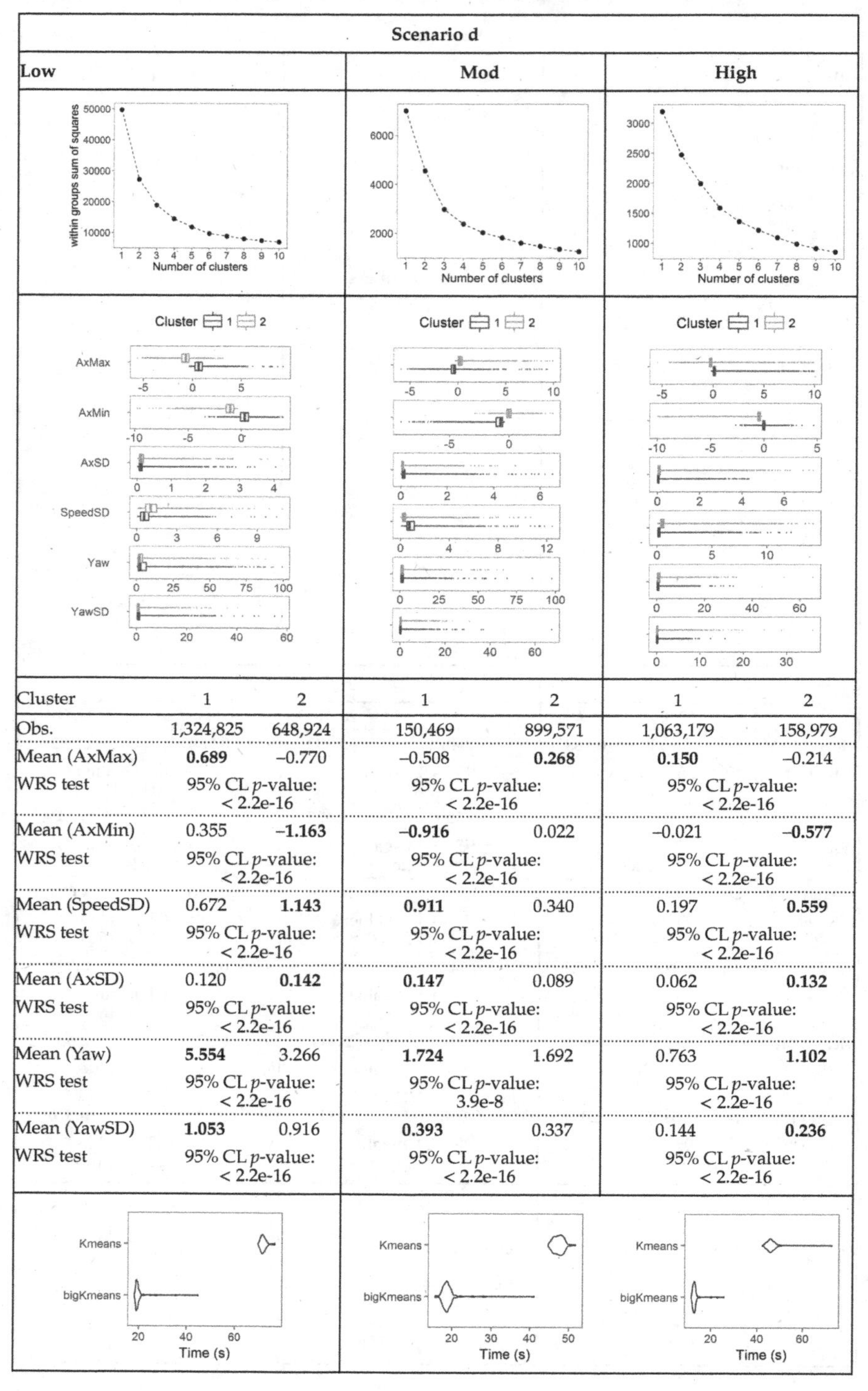

Scenario d						
	Low		Mod		High	
Cluster	1	2	1	2	1	2
Obs.	1,324,825	648,924	150,469	899,571	1,063,179	158,979
Mean (AxMax)	**0.689**	–0.770	–0.508	**0.268**	**0.150**	–0.214
WRS test	95% CL *p*-value: < 2.2e-16		95% CL *p*-value: < 2.2e-16		95% CL *p*-value: < 2.2e-16	
Mean (AxMin)	0.355	**–1.163**	**–0.916**	0.022	–0.021	**–0.577**
WRS test	95% CL *p*-value: < 2.2e-16		95% CL *p*-value: < 2.2e-16		95% CL *p*-value: < 2.2e-16	
Mean (SpeedSD)	0.672	**1.143**	**0.911**	0.340	0.197	**0.559**
WRS test	95% CL *p*-value: < 2.2e-16		95% CL *p*-value: < 2.2e-16		95% CL *p*-value: < 2.2e-16	
Mean (AxSD)	0.120	**0.142**	**0.147**	0.089	0.062	**0.132**
WRS test	95% CL *p*-value: < 2.2e-16		95% CL *p*-value: < 2.2e-16		95% CL *p*-value: < 2.2e-16	
Mean (Yaw)	**5.554**	3.266	**1.724**	1.692	0.763	**1.102**
WRS test	95% CL *p*-value: < 2.2e-16		95% CL *p*-value: 3.9e-8		95% CL *p*-value: < 2.2e-16	
Mean (YawSD)	**1.053**	0.916	**0.393**	0.337	0.144	**0.236**
WRS test	95% CL *p*-value: < 2.2e-16		95% CL *p*-value: < 2.2e-16		95% CL *p*-value: < 2.2e-16	

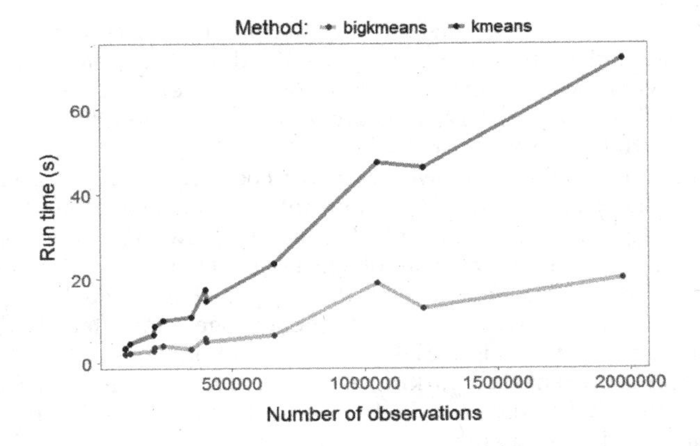

Figure 7.2 Data processing time comparison between two implementations of K-means.

7.4.2 Hot Spot Identification

A critical issue in traffic safety is to identify high-risk locations for better safety planning. Data visualization was used to show the spatial distribution of aggressive driving based on the results of scenario a. Figure 7.3 illustrates where most aggressive driving occurred in Ann Arbor, Michigan. As expected, aggressive

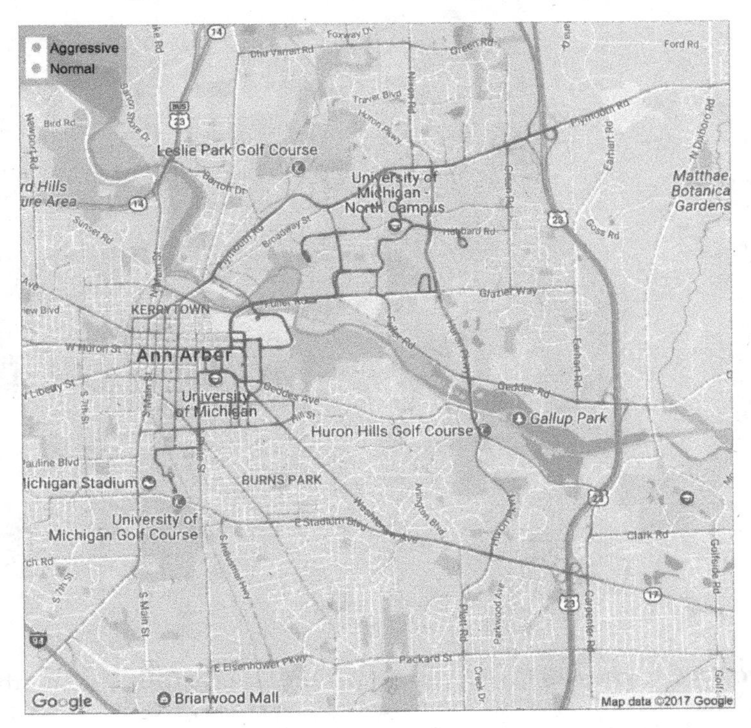

Figure 7.3 Visualization of aggressive vs. normal driving in Ann Arbor (slightly zoomed in for better clarity).

driving mostly occurred near intersections, ramps, and horizontal curves where drivers encounter changes in roadway geometric design or need to interact with other drivers as well as infrastructure control systems. Color intensity shows the relative frequency of aggressive driving events and thus a location in orange color with higher color intensity means more aggressive driving events compared to orange color with lower intensity. Focusing on the dense parts of Ann Arbor as shown in Figure 7.4a, various intersections experienced different frequency of aggressive driving events. Figure 7.4b shows a three-legged signalized intersection with a high number of aggressive driving events concentrated on only one approach. Figure 7.4c illustrates a relatively long track of aggressive events and it also shows two spots with higher orange color intensity that suggests further investigation. Figure 7.4d shows a location where a high number of aggressive events presents at a parking entrance. This analysis could be further detailed to assist local traffic agencies in prioritizing resource distribution and countermeasures development.

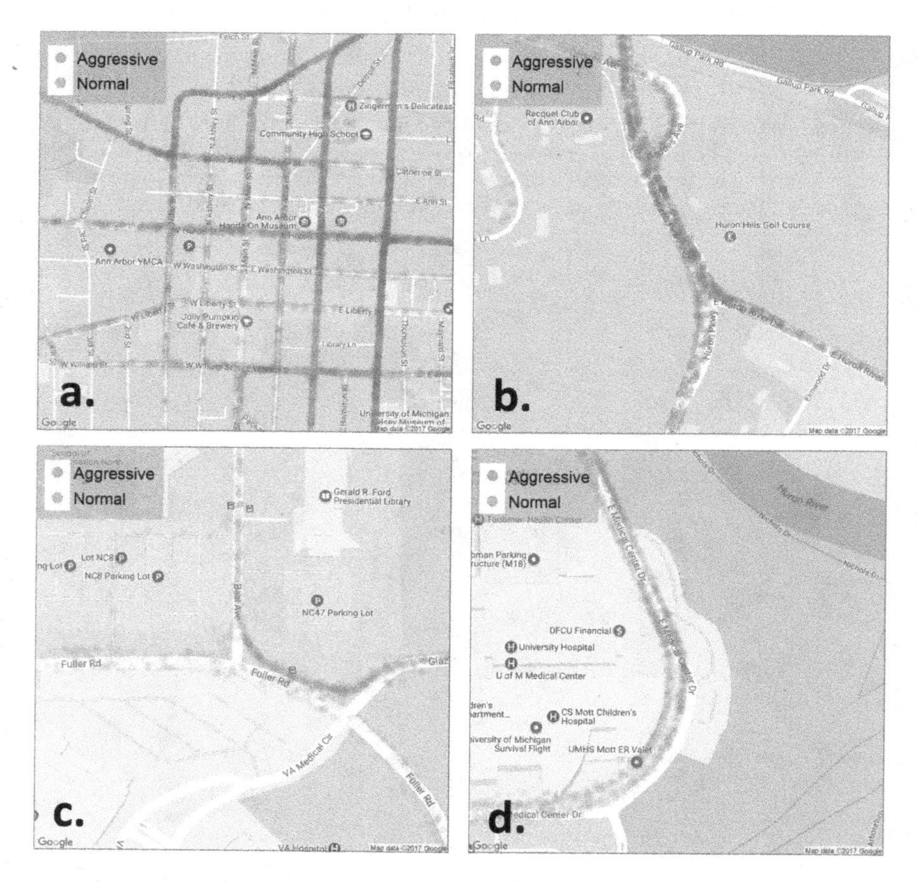

Figure 7.4 Aggressive driving visualization in different parts of Ann Arbor city: (a) intersections experienced different frequency of aggressive driving events, (b) three-legged signalized intersection with a high number of aggressive driving events, (c) long track of aggressive events, and (d) high number of aggressive events presents at a parking entrance.

7.4.3 Impact of Trip Travel Time

This section investigates the effects of trip travel time on aggressive driving. As shown in Figure 7.5a, as trip travel time increases the number of trip segments with aggressive driving events increases too. This trend suggests that longer trips are associated with more aggressive driving which is expected; longer trips provide more opportunities to take aggressive measures. Considering a single trip, aggressive driving may occur more frequently as the travel time increases due to increase in driver fatigue, distraction, and expediency. It should be noted that there are other factors that could potentially impact the number of aggressive driving events such as level of traffic, driver emotional status, etc. However, these were not considered as no data were available to reflect these variables. Also, as previously described, some trip segments with invalid data were removed resulting in not having continuous trajectory data for all trips. Therefore, trips having less than 95% of trip data were excluded in this analysis resulting in 692 trips with average trip travel time of 18.25 minutes.

To design the experiment, each trip was divided into two equal parts (first half vs. second half). The null hypothesis states that mean number of aggressive driving events is the same for first and second half of the trips. The alternative hypothesis states that more aggressive driving events occur in the second half comparing to the first half. This hypothesis testing was conducted for three travel time categories of short (5–15 minutes), moderate (15–30 minutes), and long (>30 minutes) separately. Similar to variable comparison as discussed earlier, a non-parametric test, namely the WRS test was employed, but in this case the paired test was appropriate since we had pairs of observations. The comparison is visualized in Figure 7.5b. The small *p*-value (2.1e-6) for moderate trips rejected the null hypothesis suggesting that more aggressive driving is likely to occur in the second half of trips with 95% confidence level. However, there is no strong evidence to reject the null hypothesis for short and long trips as high *p*-values of 0.48 and 0.22 were resulted, respectively. Intuitively, when travel time is short, it is not expected that the number of aggressive driving events would change for the second half of the trip comparing to the first half since the trip may not be long enough to produce fatigue or distraction. For long trips, although the mean number of aggressive driving events is higher in the second half, this increase is not statistically significant. The reason might be due to small number of observations for this travel time category compared to other categories (92 in long vs. 370 in short and 230 in moderate categories).

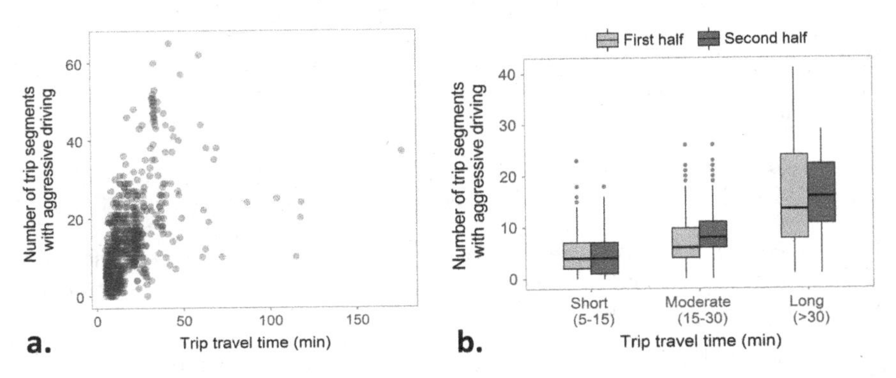

Figure 7.5 Impact of travel time on aggressive driving for (a) overall travel time and (b) travel time categories of short, moderate, and long.

7.5 CONCLUSIONS

This study applied an unsupervised learning method, namely K-means, to distinguish aggressive driving from normal driving based on kinetic data that were obtained as part of the SPMD study. Two model implementations were used to show the impacts on data processing time. It was shown that the big K-means implementation is significantly more efficient than the regular kmeans implementation in R statistical environment. Also, a chunk-wise data storing and processing approach was adopted to handle the large dataset on a personal computer, which allowed processing of more than 80 million rows of data. Dividing trip segments into two clusters of aggressive and normal was conducted for different trip segment lengths. For shorter trip segments, a clear distinction of clusters between aggressive and normal based on observed extreme values for several variables was not possible as each variable may indicate a specific aggressive measure (extreme yaw may indicate an abrupt lane change vs. extreme deceleration may indicate hard braking). For longer segments, it was possible to observe extreme values of all variables occurring in one cluster. Thus, for trip segments of 10 seconds, aggressive driving events were identified based on the observed extreme values for all variables. These events were also visualized throughout the city of Ann Arbor, Michigan to illustrate how a safety planning tool can be developed bead on the clustering model results. In addition, it was found based on the dataset used that trips with moderate travel time (15–30 minutes) experience more aggressive driving events in their second half comparing to the first half. No evidence supported this phenomenon for trips with short or long travel time. Moreover, more observations could help to better generalize the hypothesis that drivers are more likely to take aggressive actions in the second half of trips than the first half. The future work focuses on developing clustering models with higher number of clusters for shorter trip segments to represent different types of aggressive behavior. Future work will also take a personalized aggressive behavior approach where individual drivers are evaluated instead of the entire pool of drivers in the dataset.

REFERENCES

G. Abe and J. Richardson, "Alarm timing, trust and driver expectation for forward collision warning systems," *Appl. Ergon.*, vol. 37, no. 5, pp. 577–586, Sep. 2006.

American Automobile Association, "Aggressive driving." Online. Available: www.aaafoundation.org/aggressive-driving. Accessed: 21 Oct 2017.

S.-W. Chen, C.-Y. Fang, and C.-T. Tien, "Driving behaviour modelling system based on graph construction," *Transp. Res. Part C Emerg. Technol.*, vol. 26, no. Supplement C, pp. 314–330, Jan. 2013.

S. T. Chrysler, J. M. Cooper, and D. C. Marshall, "Cost of warning of unseen threats," *Transp. Res. Rec. J. Transp. Res. Board*, vol. 2518, pp. 79–85, Jan. 2015.

I. De Vlieger, D. De Keukeleere, and J. G. Kretzschmar, "Environmental effects of driving behaviour and congestion related to passenger cars," *Atmos. Environ.*, vol. 34, no. 27, pp. 4649–4655, Jan. 2000.

S. Doecke, A. Grant, and R. W. G. Anderson, "The real-world safety potential of connected vehicle technology," *Traffic Inj. Prev.*, vol. 16, Supplement 1, pp. S31–S35, 2015.

J. W. Emerson and M. J. Kane, "biganalytics: Utilities for 'big.matrix' Objects from Package 'bigmemory'. R package version 1.1.14. https://CRAN.R-project.org/package=biganalytics," 2016.

Federal Highway Administration, "Research data exchange," 2015. Online. Available: www.its-rde.net/home. Accessed: 17 Jul 2016.

W. Genders and S. N. Razavi, "Impact of connected vehicle on work zone network safety through dynamic route guidance," *J. Comput. Civ. Eng.*, vol. 30, no. 2, Mar. 2016.

S. Ghanipoor Machiani and M. Abbas, "Assessment of driver stopping prediction models before and after the onset of yellow using two driving simulator datasets," *Accid. Anal. Prev.*, vol. 96, pp. 308–315, 2016.

S. Ghanipoor Machiani, A. Jahangiri, V. Balali, and C. Belt, "Predicting driver risky behavior for curve speed warning systems using real field connected vehicle data," *Presented at the Transportation Research Board 96th Annual Meeting Transportation Research Board*, Washington, DC, 2017.

N. Goodall, B. Smith, and B. Park, *Microscopic Estimation of Freeway Vehicle Positions Using Mobile Sensors*. 91st Annual Meeting of the Transportation Research Board, Washington, DC. 2012.

F. Guo and Y. Fang, "Individual driver risk assessment using naturalistic driving data," *Accid. Anal. Prev.*, vol. 61, pp. 3–9, Dec. 2013.

M. Haglund and L. Åberg, "Speed choice in relation to speed limit and influences from other drivers," *Transp. Res. Part F Traffic Psychol. Behav.*, vol. 3, no. 1, pp. 39–51, Mar. 2000.

M. Ishibashi, M. Okuwa, S. 'ichi Doi, and M. Akamatsu, "Indices for characterizing driving style and their relevance to car following behavior," in *SICE Annual Conference 2007*, 2007, Kagawa University Takamatsu, Kagawa, Japan, pp. 1132–1137.

A. Jahangiri, V. Berardi, and S. Ghanipoor Machiani, "Application of real field connected vehicle data for aggressive driving identification on horizontal curves," *IEEE Intell. Transp. Syst. Trans*, 19(7), pp. 2316–2324, July 2018. doi: 10.1109/TITS.2017.2768527.

A. Jahangiri, M. Elhenawy, H. Rakha, and T. A. Dingus, "Investigating cyclist violations at signal-controlled intersections using naturalistic cycling data," in *Intelligent Transportation Systems (ITSC), 2016 IEEE 19th International Conference on*, 2016, Rio de Janeiro, Brazil, pp. 2619–2624.

A. Jahangiri, H. Rakha, and T. A. Dingus, "Predicting red-light running violations at signalized intersections using machine learning techniques," *Presented at the Transportation Research Board 94th Annual Meeting*, Washington, D.C., 2015.

A. Jahangiri, H. Rakha, and T. A. Dingus, "Red-light running violation prediction using observational and simulator data," *Accid. Anal. Prev.*, 2016, 96, pp. 316–328.

A. K. Jain, "Data clustering: 50 years beyond K-means," *Pattern Recognit. Lett.* 2010, 31(8), pp. 651–666.

M. Kamrani, B. Wali, and A. J. Khattak, "Can data generated by connected vehicles enhance safety? A proactive approach to intersection safety management," *ArXiv Prepr. ArXiv170900743*, Transportation Research Record: Journal of the Transportation Research Board 2659, 2017, 80–90.

D. J. Ketchen Jr and C. L. Shook, "The application of cluster analysis in strategic management research: an analysis and critique," *Strateg. Manag. J.*, 1996, pp. 441–458.

E. Kim and E. Choi, "Estimates of critical values of aggressive acceleration from a viewpoint of fuel consumption and emissions," in *2013 Transportation Research Board Annual Meeting*, Washington, D.C., 2013, no. 13–3443.

R. Knipling, L.N. Boyle, J. S. Hickman, J. S. York, and E. C. Olsen, "Individual differences and the 'high-risk' commercial driver: A synthesis of safety practice," *Presented at the Transportation Research Board*, Washington, DC, 2004.

B. Lantz, *Machine Learning with R*. Packt Publishing Ltd, 2015.

S. Li *et al.*, "Geospatial big data handling theory and methods: a review and research challenges," *ISPRS J. Photogramm. Remote Sens.*, vol. 115, no. Supplement C, Birmingham, UK, 2016, pp. 119–133.

J. Liu and A. J. Khattak, "Delivering improved alerts, warnings, and control assistance using basic safety messages transmitted between connected vehicles," *Transp. Res. Part C Emerg. Technol.*, vol. 68, pp. 83–100, Jul. 2016.

J. Liu, A. Khattak, and X. Wang, "The role of alternative fuel vehicles: using behavioral and sensor data to model hierarchies in travel," *Transp. Res. Part C Emerg. Technol.*, vol. 55, no. Supplement C, pp. 379–392, Jun. 2015.

J. MacQueen and others, "Some methods for classification and analysis of multivariate observations," in *Proceedings of the Fifth Berkeley Symposium on Mathematical Statistics and Probability*, 1967, vol. 1, pp. 281–297.

G. A. M. Meiring and H. C. Myburgh, "A review of intelligent driving style analysis systems and related artificial intelligence algorithms," *Sensors*, vol. 15, no. 12, pp. 30653–30682, Dec. 2015.

O. A. Osman, J. Codjoe, and S. Ishak, "Impact of time-to-collision information on driving behavior in connected vehicle environments using a driving simulator test bed," *J. Traffic Logist. Eng.*, vol. 3, no. 1, 2015.

R. Paleti, N. Eluru, and C. R. Bhat, "Examining the influence of aggressive driving behavior on driver injury severity in traffic crashes," *Accid. Anal. Prev.*, vol. 42, no. 6, pp. 1839–1854, Nov. 2010.

P. J. Rousseeuw, "Silhouettes: a graphical aid to the interpretation and validation of cluster analysis," *J. Comput. Appl. Math.*, vol. 20, pp. 53–65, 1987.

Safety Pilot Model Deployment, *Sample Data, from Ann Arbor, Michigan, Version 1*. Washington DC: U.S. Department of Transportation's (USDOT) Intelligent Transportation Systems (ITS) Joint Program Office (JPO), 2014.

Safety Pilot Model Deployment, *Sample Data Environment Data Handbook, Version 1.3*. Washington, DC: U.S. Department of Transportation's (USDOT) Intelligent Transportation Systems (ITS) Joint Program Office (JPO), 2015.

F. Sagberg, Selpi, G. F. Bianchi Piccinini, and J. Engström, "A review of research on driving styles and road safety," *Hum. Factors*, vol. 57, no. 7, pp. 1248–1275, 2015.

R. Sengupta, S. Rezaei, S. E. Shladover, D. Cody, S. Dickey, and H. Krishnan, "Cooperative collision warning systems: concept definition and experimental implementation," *J. Intell. Transp. Syst.*, vol. 11, no. 3, pp. 143–155, Jul. 2007.

S. A. Soccolich, J. S. Hickman, and R. J. Hanowski, "Identifying high-risk commercial truck drivers using a naturalistic approach," National Surface Transportation Safety Center for Excellence (NSTSCE), Virginia Tech Transportation Institute, 2011.

O. Taubman-Ben-Ari, M. Mikulincer, and O. Gillath, "The multidimensional driving style inventory – scale construct and validation," *Accid. Anal. Prev.*, vol. 36, no. 3, pp. 323–332, May 2004.

R. Tibshirani, G. Walther, and T. Hastie, "Estimating the number of clusters in a data set via the gap statistic," *J. R. Stat. Soc. Ser. B Stat. Methodol.*, vol. 63, no. 2, pp. 411–423, Jan. 2001.

B. Wali, A. Ahmed, S. Iqbal, and A. Hussain, "Effectiveness of enforcement levels of speed limit and drink driving laws and associated factors – exploratory empirical analysis using a bivariate ordered probit model," *J. Traffic Transp. Eng. Engl. Ed.*, vol. 4, no. 3, pp. 272–279, Jun. 2017.

S. Walkowiak, *Big Data Analytics with R*. Birmingham Mumbai: Packt Publishing - ebooks Account, 2016.

J. Wang, Y. Zheng, X. Li, C. Yu, K. Kodaka, and K. Li, "Driving risk assessment using near-crash database through data mining of tree-based model," *Accid. Anal. Prev.*, vol. 84, pp. 54–64, Aug. 2015.

W. Wang and J. Xi, "A rapid pattern-recognition method for driving styles using clustering-based support vector machines," *2016 American Control Conference (ACC)*, Boston, MA, 2016, pp. 5270–5275.

W. Wang, J. Xi, and X. Li, "Statistical pattern recognition for driving styles based on Bayesian probability and Kernel density estimation," *ArXiv160601284 Cs Stat*, Jun. 2016.

X. Wang, A. J. Khattak, J. Liu, G. Masghati-Amoli, and S. Son, "What is the level of volatility in instantaneous driving decisions?," *Transp. Res. Part C Emerg. Technol.*, vol. 58, Part B, pp. 413–427, Sep. 2015.

G.N. Wassim, J. Koopmann, J. D. Smith, and J. Brewer, *Frequency of Target Crashes for Intellidrive Safety Systems*. Washington, DC: U.S. Dept. of Transportation, National Highway Traffic Safety Administration, 2010.

A. P. Weiss, *Psychological Principles in Automotive Driving*. Ohio State University, Columbus, OH, 1930.

Z. Yang, T. Kobayashi, and T. Katayama, "Development of an intersection collision warning system using DGPS," No. 2000-01-1301. SAE Technical Paper, 2000.

H. Yu, F. Tseng, and R. McGee, "Driving pattern identification for EV range estimation," in *2012 IEEE International Electric Vehicle Conference*, Clemson University, Greenville, SC, 2012, pp. 1–7.

8 Exploratory Analysis of Run-Off-Road Crash Patterns

Mohammad Jalayer, Huaguo Zhou, and Subasish Das

CONTENTS

8.1 INTRODUCTION

Roadway departure (RwD) occurs when a vehicle departs from the traveled way by crossing an edgeline or a centerline (FHWAa n.d.a). RwD events comprise both run-off-road (ROR) and cross-median/centerline head-on collisions. Most head-on crashes are similar to ROR crashes – in both cases, the vehicle strays from its travel lane (Neuman et al. 2003a,b). The vast majority of RwD events are generally identified in fatality analysis reporting system (FARS) by finding crashes in which the first sequence of event for any vehicle involved in the crash is one of the following: ran off road-right, ran off road-left, cross-median, or cross-centerline (Satterfield 2014). Factors contributing to ROR collisions can be divided into two major categories: infrastructure and environmental factors and driver factors. Examples from the first include the effect of weather on pavement conditions, travel lanes that are too narrow or have substandard curves, and unforgiving roadsides. Driver factors include traveling too fast through a curve or down a grade; a driver attempting to avoid a vehicle, an object, or an animal in the travel lane; and inattentive driving due to distraction, fatigue, sleep, or drugs (Neuman et al. 2003a,b). Compared to other crash types, RwD is one of the most severe types of crashes (FHWA n.d.a; Jalayer and Zhou 2016a). An analysis of statistics from the FARS database for crash data from 2007 to 2013 reveal that an average of 59 percent of annual motor vehicle traffic fatalities in the United States occurred due to RwD (NHTSA 2016). Moreover, according to the FHWA, 80 percent of total ROR fatalities occurred on rural highways, and about 90 percent of those occurred on two-lane roads (Lord et al. 2011; FHWA n.d.b), the roadway type upon which this paper is focused.

To determine the most significant contributing factors, and then develop effective safety countermeasures, these numbers require further analysis. A major challenge for state and local agencies is to find patterns in these huge databases. Exploratory data analysis (EDA) is an approach by which patterns, changes, and anomalies in large datasets may be determined, beyond the hypothesis testing task or formal modeling (Cook and Swayne 2007; Chatfield 1995). Using a variety of mostly graphical techniques (e.g., box plot, scatter plot, multiple correspondence analysis, and principal component analysis), EDA can extract specific information from datasets and transform it into an understandable structure. Since ROR crashes accounted for the majority of RwD events (about 80 percent), this study uses multiple correspondence analysis (MCA) to identify the key factors contributing to ROR collisions related to the roadway and roadside geometric design features of rural two-lane roads. The MCA method identifies patterns in complex datasets and measures significant contributing factors and their degree of association. To employ this method, datasets from the United States Road Assessment Program (usRAP), a program of the American Automobile Association Foundation for Traffic Safety, were obtained and 5 years (2009–2013) of ROR crash data in Illinois were gathered. To achieve the program's Toward Zero Deaths vision, agencies are working to decrease the frequency and severity of RwD crashes. The results of this study can help researchers and transportation agencies to get a better knowledge of the major contributing factors to ROR crashes and prioritize the locations where safety countermeasures should be implemented (e.g., signage, pavement safety measures, and roadside design improvements).

8.2 LITERATURE REVIEW

There have been a considerable number of studies identifying various contributing factors to ROR crashes, using a variety of data collection and data analysis methods. In an attempt to identify contributing factors to ROR crashes, McLaughlin et al. (McLaughlin et al. 2009) obtained the dataset from a 100-car naturalistic driving study. In each car, seven various software and hardware instruments had been installed to collect data. In the study, an ROR event was identified as having occurred when the subject vehicle passed or touched a roadway boundary (e.g., edge line marking and pavement edge). The study results revealed that a single factor contributed to 75 percent of the ROR events, followed by two other factors contributing 22 percent. The analysis results showed that the most common factors contributing to ROR events included: distraction, short following distance, low friction, narrower lane, and roadside geometric configurations. Additionally, 36 percent of the ROR events involved distractions due to non-driving tasks and 30 percent of the ROR events happened on road curves. Liu and Subramanian (Liu and Subramanian 2009) evaluated various contributing factors associated with single-vehicle ROR crashes. Their results showed that horizontal road alignment, area type, speed limit, roadway geometric characteristics, and lighting conditions significantly affect the frequency and severity of ROR crashes. Lord et al. (2011) investigated the factors contributing to RwD crashes on two-way two-lane rural roads in the state of Texas. The authors divided the contributing factors into three groups, comprising highway design characteristics (i.e., lane width, shoulder width and type, roadside design, pavement edge drop-off, horizontal curvature and grades, driveway and pavement surfaces, and traffic volume), human factors (i.e., speeding, alcohol and drug use, and age and gender), and other factors (i.e., time of day, vehicle type). The data for the crashes, geometric road characteristics, bridges and curves, and traffic characteristics were gathered from various databases and then combined. The results demonstrated that, compared to tangent sections, wider shoulders yielded greater safety effects on horizontal curves. Additionally, most RwD crashes occurred on weekends, being attributed to people driving under the influence (DUI). Unlike driveway density, which had a little impact on RwD crashes, lighting conditions had a great influence on the probability of an RwD crash occurrence.

In another study conducted by the National Highway Traffic Safety Administration, driver inattention, driver fatigue, roadway surface conditions, driver blood alcohol presence, drivers' level of familiarity with the roadway, and driver gender were identified as the most significant factors contributing to ROR crashes (Liu and Ye 2011). Jalayer and Zhou (2016b) presented a new approach to evaluating the safety risk of roadside features for rural two-lane roads based on reliability analysis. The authors confirmed that reliability indices could serve as indicators to gauge safety levels. Eustace et al. (2014) identified the most significant factors contributing to severe ROR crashes (i.e., injury and fatal) using generalized ordered logit regression. Their results demonstrated that driver conditions (e.g., impaired drivers), road alignments (e.g., curves), roadway characteristics (e.g., grade), gender (e.g., male), and roadway surface conditions (e.g., wet) increased the likelihood of severe ROR crashes. In an attempt to determine unforgiving roadside contributing factors, Roque et al. (2015) collected ROR crash data on freeway road sections in Portugal and developed multinomial and mixed logit regression models, accordingly. The empirical findings of this study indicated that critical slopes and horizontal curves significantly contributed toward the fatal ROR crashes. In 2015, the American Traffic Safety Services Association published a booklet as an executive summary of various case studies to educate transportation practitioners regarding ROR crashes and associated safety countermeasures (Jalayer and Zhou 2016; ATSSA 2015). In this booklet, countermeasures are categorized as signs (e.g., chevron), pavement safety (e.g., high friction surface treatments), and roadside design (e.g., clear zone improvements). The results of this study found pavement safety countermeasures, compared to other categories, to be the most effective in reducing total ROR crash frequency and severity.

Regarding the methodology outlined in this study, while there is an extensive body of literature on the application of statistical modeling in transportation science (Soltani-Sobh et al 2015; Khalilikhah and Heaslip 2016a,b; Sharifi and Shabaniverki 2016; Heaslip et al. 2014; Sharifi et al. 2015; Pour-Rouholamin and Jalayer 2016; Baratian-Ghorghi et al. 2015, 2016), few past studies have applied MCA to highway safety. Das and Sun (2015) employed MCA to analyze 8 years' worth of vehicle-pedestrian crash data in Louisiana. In another study, Das and Sun (2016) applied MCA method to determine the contributing factors in fatal ROR crashes using 8 years (2004–2011) of Louisiana crash data. Using MCA, Factor et al. (2010) investigated the social morphology of car accidents over a 20-year period. Nallet et al. (2008) employed MCA to identify the effect of driving license points recovery courses on attending drivers' road crashes. Kim and Yamashita (2008) also used MCA to explore the characteristics of pedestrian-involved collisions in Hawaii. In another study, Fontaine (1995) applied MCA method to analyze the topology of vehicle-pedestrian accidents.

It should be noted that although there are a considerable number of studies of the factors contributing to ROR crashes (Lee and Mannering 1999, 2002; McGinnis et al. 2001; Wu et al. 2004; Spainhour et al. 2008; Motella 2009; De Albuquerque et al. 2010; Savolainen 2011; Peng and Boyle 2012; Fitzpatrick 2013; Hallmark et al. 2013; Jalayer et al. 2015; Dissanayake and Roy 2014; Gates et al. 2014; Petegem and Wegman 2014; Jalayer and Zhou 2013; Schrum et al. 2014; Hallmark et al. 2015; Roque and Cardoso 2015), very few have used graphical EDA techniques for crash analysis. To our knowledge, no previous analyses of the usRAP database have investigated the effects of roadway and roadside geometric design features on ROR crash frequency and severity, which we address in this paper.

8.3 METHOD AND DATA

8.3.1 Multiple Correspondence Analysis

MCA, as an increasingly popular EDA technique, is a powerful method for analyzing and graphically presenting the relationship patterns among several categorical (nominal-scale) dependent variables in large and complex datasets. MCA is able to interpret the large datasets without the necessity of any preconditions (Das and Sun 2015; Nallet 2008; Panagiotakos and Pitsavos 2004). Moreover, in both of count data models and crash severity models the sample size significantly influences model performance (Ye and Lord 2014). MCA's graphical overviews simplify the expression of the relationships between variables, thereby making interpretation easier (Das and Sun 2016; Abdi and Valentin 2007; Greenacre and Blasius 2006). More detailed descriptions of the MCA method and its development history can be found in Greenacre and Blasius (2006), Gifi (1991), and Le Roux and Rouanet (2010).

Denoting I as the set of i individual records and J as the set of categories of all variables, MCA is performed on an $I \times J$ design or indicator matrix (Das and Sun 2015; LeRoux and Rouanet 2010). Therefore, an entry in the cell (i, j) includes the individual record i and category j. For instance, gender is one nominal variable with two values, male vs. female, corresponding "0 1" for the male and "1 0" for the female. Accordingly, the completed matrix includes the binary columns with one and only one column, per nominal variable, which takes the value of "1." The categories can be either qualitative or represent the outcome of the splitting of quantitative variables into categories. In MCA, associated categories are placed close together in a Euclidean space, leading clouds, or combinations of points that have similar distributions (Das and Sun 2015; LeRoux and Rouanet 2010). Notably, MCA produces two-point clouds, including an individual records cloud and a categories cloud, which are defined by one-, two-, or three-dimensional graphs (Das and Sun 2015). It should be noted that the distances between points within a variable in the N-dimensional graph are summaries of all the information about the similarities between all the individual records (Das and Sun 2015; Greenacre and Blasius 2006). Since a lower-dimensional space that includes all or nearly all of the information is desirable, especially for large and complex databases, the two-dimensional graph is the most convenient, with its illustrative planar surface. The fundamental principles of the two types of point clouds are described in the following sections.

8.3.1.1 Cloud of Individuals

As mentioned above, the construction of clouds is based on the set of all distances between individual records for a variable in the database, for which different categories have been selected. In other words, if two individual records i and i' select the same category for variable m, the distance between them will be zero. Otherwise, for each variable, the squared distance between individuals associated with each category is calculated based on Equation 8.1 (Das and Sun 2015, 2016; LeRoux and Rouanet 2010):

$$d_m^2(i,i') = \frac{1}{f_j} + \frac{1}{f_{j'}} \tag{8.1}$$

where

$d_m^2(i,i')$ is the squared distance between individuals i and i' for variable m
f_j is the relative frequency of individual records that selected category j
$f_{j'}$ relative frequency of individual records that selected category j'

For variable "lane width," as an example, the individual records i and $i'$are two different roadway segments and the categories j and j' are two different lane widths (e.g., "less than 9 ft." and "9–10.6 ft."). The relative selection frequency of each category is defined as the total number of individual records that chose that particular category divided by the total number of individual records (n) in the database [27, 28]. In order to obtain the overall squared distance between two individual records i and i', all individual squared distances must be added together, as shown in Equation 8.2 [27,57]:

$$D^2(i,i') = \frac{1}{M}\sum_{m \in M} d_m^2(i,i') \tag{8.2}$$

where

$D^2(i,i')$ is the overall squared distance between individuals i and i'
$d_m^2(i,i')$ is the squared distance between individuals i and i' for variable m
M is the set of all variables

8.3.1.2 *Cloud of Categories*

The cloud of categories has the same dimension as the cloud of individuals. Category j is defined by a point, namely N^j, with weight (n_j), which is the number of individuals that selected this category. The squared distance between categories j and j' can be written as in Equation 8.3 (Das and Sun 2015; LeRoux and Rouanet 2010):

$$(N^jN^{j'})^2 = \frac{n_j + n_{j'} - 2n_{jj'}}{n_j n_{j'} / n} \tag{8.3}$$

where

$(N^jN^{j'})^2$ is the squared distance between categories j and j'
n_j is the number of individuals that selected category j
$n_{j'}$ number of individuals that selected category j'
$n_{jj'}$ is the number of individuals that selected both categories j and j'

8.3.2 Data Collection

In order to evaluate the proposed MCA method, the required data for crashes and roadway/roadside geometric features from two databases were gathered and combined. The historical ROR crash data for a 5-year period from 2009 through 2013 were compiled from the Illinois Department of Transportation (IDOT) (Zhou et al. 2013). The roadway/roadside geometric design features of 4,500 300-ft roadway segments were also gathered from the usRAP database in Illinois (usRAP 2013). The usRAP database is an efficient tool that provides information in accessible formats regarding crash risk from the viewpoints of public and highway agencies. In the pilot program, the eight participating states included Florida, Illinois, Iowa, Kentucky, Michigan, New Jersey, New Mexico, and Utah (OST-R n.d.). The usRAP database contains data about roadways, roadsides, and bicycle and pedestrian facilities, all of which contribute to vehicle crashes.

For the purposes of this study, a set of key variables for further investigation from among all the parameters included in the usRAP database was nominated, based on engineering study results gleaned from a comprehensive literature review. These variables comprise roadside severity, paved and unpaved shoulder widths, lane width, shoulder rumble strips, horizontal curvature, delineation condition, vertical alignment variation, road condition, land use, and speed limit. Table 8.1 lists all the contributing variables and categories, along with their frequencies and percentages. According to Table 8.1, for some variables, the majority of segments fall into one or two categories. For instance, more than 89 percent of segments have no horizontal curvature, 98 percent of segments are without shoulder rumble strips, and 95 percent have good road conditions. Roadside severity indicates the nature of and/or distance to the nearest roadside object, which could result in a fatal or serious injury to vehicle occupants (usRAP 2012). Since only segments with the same annual average daily traffic (AADT) range, 6,500–7,500 vehicles per day were considered, in this study, the effect of AADT on ROR crashes could not be considered. It should be noted that the fixed segmentation rule (i.e., 300 ft.) and thresholds for roadway characteristics such as lane width and shoulder width, in this study, are defined by the usRAP database (usRAP 2012). Moreover, the "crash severity" variable is corresponding to the most severe crash occurring on a segment in cases multiple crashes occurred on a segment. Figure 8.1 illustrates a flowchart of the working steps used in this study.

8.4 RESULTS AND DISCUSSIONS

To analyze the dataset and plot the two-dimensional graphs, *R Version 3.02* statistical software and the *FactoMineR* package were employed. In a two-dimensional graphical display with two principal axes, associated categories are close together and form the point clouds (Das and Sun 2015, 2016; LeRoux and Rouanet 2010). The output is the magnitude of information associated with each dimension, which is given a value between 0 and 1, known as the eigenvalue. The eigenvalue of each dimension can serve as a dependable indicator of the total variance among variables (Das and Sun 2016). The eigenvalues of the first, second, and third dimensions are higher than others, so a two-dimensional graph carries most, but not all, the information. Figure 8.2 illustrates the percentages of variance for the top five dimensions.

Based on Figure 8.2, the first and second principal axes on the planar surface, the MCA plot, describes 10.9 percent of the total variances together. The low eigenvalues demonstrate that the variables in the database are heterogeneous due to the random nature of road segments characteristics and occurrence of accidents (Das and Sun 2015, 2016). Every point on each plot has its own coordination for all dimensions, and, obviously, the scale of the plot is highly dependent on the volume of contributions of each dimension. Figure 8.3 depicts all the study variables and their relative proximity on the map.

Regarding the interpretation of the MCA plots, it should be noted that similar objects can be compared based on their relative distances on the graph. In other words, individual records, variables, and categories within a variable may be compared just by looking at the distance between the points on the map (Das and Sun 2015; Gifi 1991; LeRoux and Rouanet 2010). As for non-similar objects, such as categories of different variables, an imaginary line from each point of interest to the centroid of the map must be considered, and then the angle defined between those lines. A very small angle indicates a relatively strong relationship, and a right angle shows that there is no association between those particular objects. An angle of more than 90 degrees denotes a negative association (Correspondence Analysis, 2016). Figure 8.3 shows that many variables are located closely to each other, thus making the same contribution to all the variances. The closer a point is located to the centroid in one dimension, the less

Table 8.1: Distributions of Segments Based on Study Categories

Variable	Category	Frequency	Percentage (%)	Description
Speed limit (mph)	Less than 40	243	5.4	
	40 to 50	1733	38.5	
	Greater than 50	2524	56.1	
Lane width (ft.)	Less than 9	63	1.4	
	9 to 10.6	1422	31.6	
	Greater than 10.6	3015	67.0	
Paved shoulder width (ft.)	Less than 3	2092	46.5	
	3 to 7.9	374	43.3	
	Greater than 7.9	86	1.9	
	None (No Shoulder)	1948	43.3	
Unpaved shoulder width (ft.)	Less than 3	2340	52.0	
	3 to 7.9	1471	32.7	
	Greater than 7.9	36	0.8	
	None (No shoulder)	653	14.5	
Shoulder rumble strips	Present	95	2.1	
	Not present	4405	97.9	
Horizontal curvature	No curvature	4014	89.2	
	Moderate curve	459	10.2	Can be driven at a maximum speed of 45 to 60 mph
	Sharp curve	27	0.6	Can be driven at a maximum speed of 25 to 45 mph
Delineation	Adequate	4428	98.4	Warning signs and pavement marking are generally presented
	Poor	72	1.6	Warning signs and pavement marking are absent or faded
Vertical alignment	Level or constant grade	4338	96.4	No roadway gradient or constant grade
Variation	Rollins	162	3.6	Moderate changes in road gradient
Road condition	Good	4279	95.1	Very few to no deficiencies
	Medium	194	4.3	A number of minor deficiencies
	Poor	27	0.6	Substantial deficiencies
Land use-left side	Undeveloped	4018	89.3	
	Residential	315	7.0	
	Commercial	45	1.0	
	Developed other than residential or commercial	122	2.7	

(*Continued*)

Table 8.1 (Continued): Distributions of Segments Based on Study Categories

Variable	Category	Frequency	Percentage (%)	Description
Land use-right side	Undeveloped	3982	88.5	
	Residential	311	6.9	
	Commercial	27	0.6	
	Developed other than residential or commercial	180	4.0	
Roadside severity - left side	Traffic barrier	225	5.0	
	Cut	9	0.2	
	Deep drainage ditches	23	0.5	
	Steep fill embankment slopes	225	5.0	
	Distance to objects 0 to 15 ft.	486	10.8	
	Distance to objects 15 to 30 ft.	2240	49.8	
	Distance to objects greater than 30 ft.	1260	28.0	
	Cliff	32	0.7	
Roadside severity - right side	Traffic barrier	162	3.6	
	Cut	0	0.0	
	Deep drainage ditches	27	0.6	
	Steep fill embankment slopes	189	4.2	
	Distance to objects 0 to 15 ft.	378	8.4	
	Distance to objects 15 to 30 ft.	2250	50.0	
	Distance to objects greater than 30 ft.	1467	32.6	
	Cliff	27	0.6	
Crash severity	Fatal	99	2.2	
	Injury	396	8.8	
	Property Damage Only (PDO)	707	15.7	
	None (No Crash)	3298	73.3	
Number of ROR crashes per segment	Zero (No Crash)	3298	73.3	
	One	895	19.9	
	Two	180	4.0	
	Three	68	1.5	
	Four	32	0.7	
	Five	27	0.6	

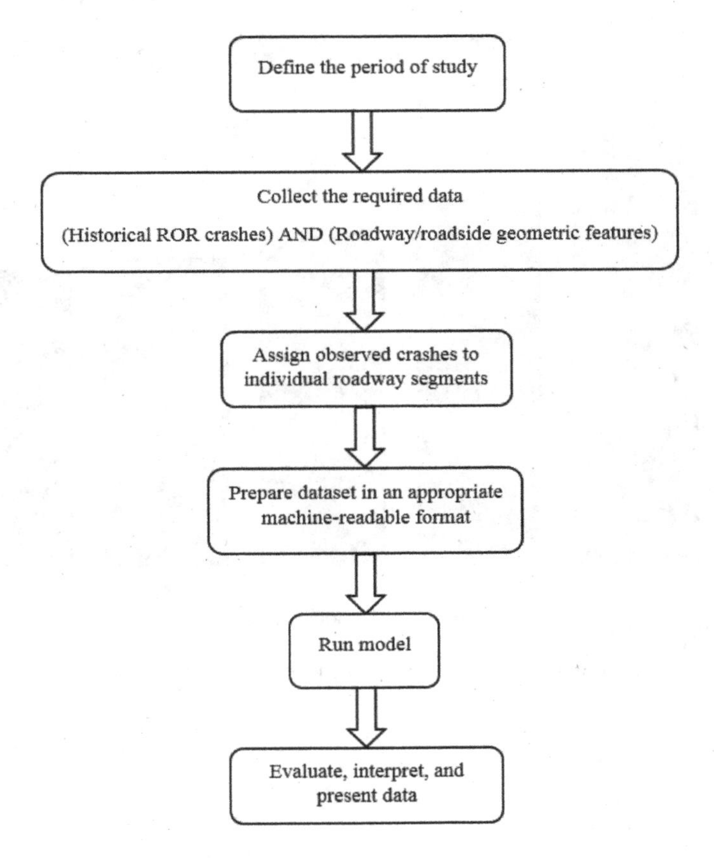

Figure 8.1 A flowchart of working steps.

it contributes to the eigenvalue of that particular dimension, making it as a relatively less important variable (Das and Sun 2016). Therefore, for dimension one, the roadside severity of the right side, the roadside severity of the left side, and the horizontal curvature contributed the most. Similarly, for dimension two, the roadside severity of the left side, the roadside severity of the right side, and the paved shoulder width are the most significant variables on ROR crash frequency and severity. Table 8.2 lists in descending order of significance all the ROR contributing factors in this study, considering the coefficient of determination (R^2) and a p-value of the overall test (F-test). R^2 ranges from 0 to 1, with 0 being no *relationship* and 1 being a *very strong relationship* between the qualitative variable and the MCA dimension (Das and Sun 2015; LeRoux and Rouanet 2010). As it can be seen, compared to roadside severity, the risk of ROR crashes is not strongly associated with delineation, land use, or vertical alignment variation.

Figure 8.4 shows the top 20 categories that contributed the most to the two-dimensional plot. Based on the relative proximity of points, several point clouds for categories can be created. According to this figure, one combination cloud correlates five ROR crashes with sharp horizontal curvature and the presence of a cliff as a roadside condition. This means that segments with a severe roadside condition and horizontal curvature are associated with a significant increase

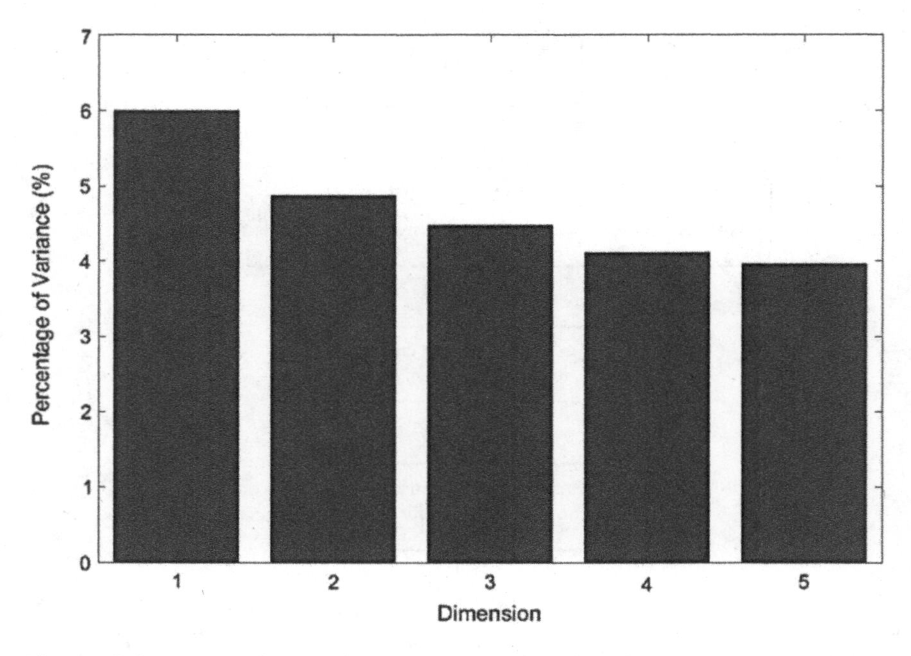

Figure 8.2 Eigenvalues and variances of the top five dimensions.

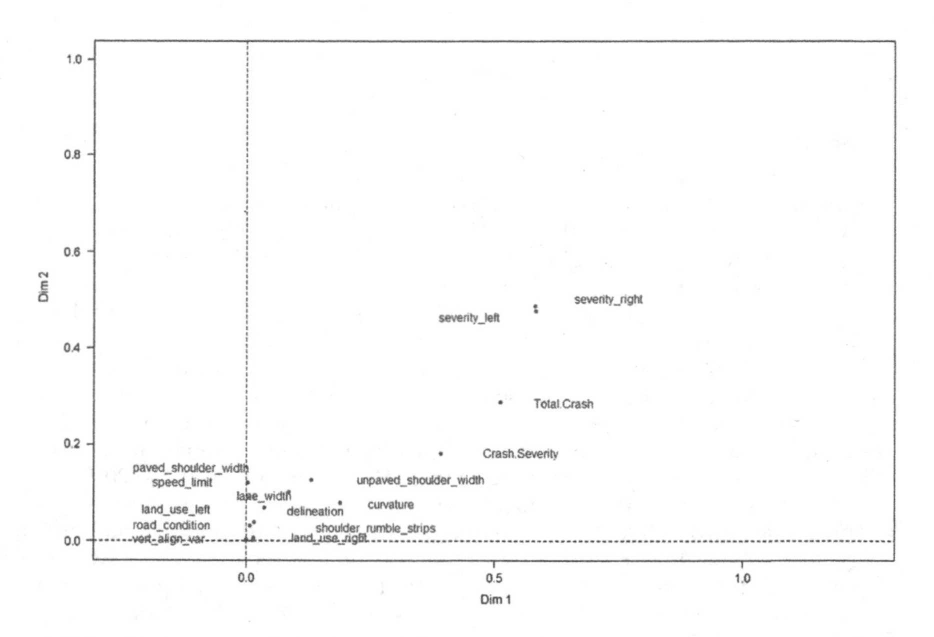

Figure 8.3 MCA plot of all study variables.

Table 8.2: Significance of Test Results for Key ROR Contributing Factors in Top Two Dimensions

	Variable	R^2	*p*-value
Dimension 1	Roadside severity-right side	0.584	<0.001
	Roadside severity-left side	0.582	<0.001
	Total crash	0.512	<0.001
	Crash severity	0.393	<0.001
	Horizontal curvature	0.191	<0.001
	Paved shoulder width	0.131	<0.001
	Unpaved shoulder width	0.088	<0.001
	Lane width	0.037	<0.001
	Delineation	0.017	<0.001
	Road condition	0.015	0.002
	Shoulder rumble strips	0.013	0.003
Dimension 2	Roadside severity-left side	0.487	<0.001
	Roadside severity-right side	0.476	<0.001
	Total crash	0.288	<0.001
	Crash severity	0.182	<0.001
	Paved shoulder width	0.127	<0.001
	Speed limit	0.121	<0.001
	Unpaved shoulder width	0.102	<0.001
	Land use-left side	0.089	<0.001
	Horizontal curvature	0.078	<0.001
	Lane width	0.068	<0.001
	Delineation	0.039	<0.001
	Land use-right side	0.031	<0.001

in the likelihood of ROR crashes, which is consistent with the findings of the majority of existing literature 14, 16, 36. Additionally, based on another cloud, most segments with three crashes had no shoulders and moderate horizontal curvature. This indicates that presence of shoulders and curves with larger radii decrease the likelihood of ROR crashes. These results are in good agreement with the findings of Roque et al. (2015), Lord et al. (2011), and Petegem and Wegman (2014). Another point cloud associated factors such as injury crashes, distance to fixed objects between 0 and 15 ft., two crashes, and the presence of a traffic barrier. Moreover, the cloud links the PDO crashes to the paved shoulder widths of 3–7.9 ft. These results are also in line with the findings of another study conducted by Lord et al. (2011).

8.5 CONCLUSION

This study utilized MCA method to identify the factors contributing to ROR crashes through combining usRAP data and historical crash records. To achieve the FHWA's *Toward Zero Deaths* vision, one of the challenges researchers and state DOTs face is how to identify key contributing factors within large and complex datasets, and then how to implement effective safety countermeasures accordingly. In conventional regression models, unlike MCA, it is required to hold the basic assumptions of regression truth and any deviation may result in incorrect outcomes (Das and Sun 2016)28. Moreover, both small and large

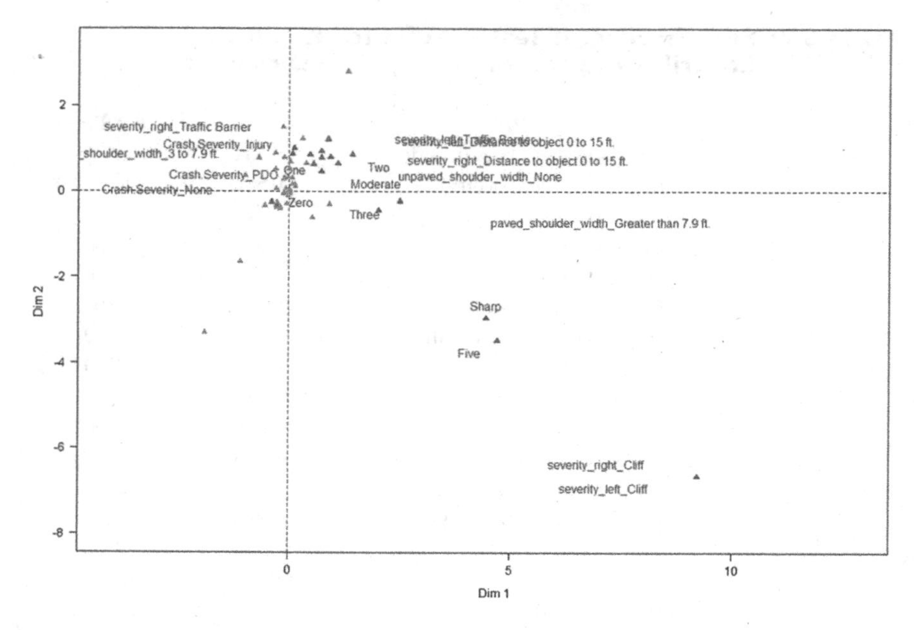

Figure 8.4 MCA plot of top 20 key categories.

sample sizes significantly influence performances of both count data models and crash severity models (Ye and Lord 2014). Since it is always possible to transform a quantitative variable into a categorical variable, and since a multi-dimensional approach to a crash will always involve a large set of categorical variables, this method is of particular interest.

To perform the model development, 5 years' worth of ROR crash data from 2009 to 2013 for the state of Illinois were obtained. In this paper, we evaluated the characteristics of the roadways and roadsides that affect ROR crashes for 4,500 300-ft segments, gathered from the usRAP database. More specifically, no previous analyses of the usRAP database have investigated the effects of roadway and roadside geometric design features on ROR crash frequency and severity. These features include roadside severity, paved and unpaved shoulder widths, lane width, shoulder rumble strips, horizontal curvature, delineation condition, vertical alignment variation, road condition, land use, and speed limit. According to the obtained results, the main contributing factors to ROR crashes are roadside severity, horizontal curvature, and shoulder width. Moreover, the likelihood of a collision with a fixed object off the road, such as s concrete barrier, is associated with increased severity of ROR crashes. Additionally, the results indicate that providing paved shoulders, with a minimum width of 3 ft., is associated with reduced ROR crash severity. It was also found that the risk of ROR crashes is not strongly associated with delineation, land use, or vertical alignment variation. One of the reasons we obtained such results is the disproportionate proportion of segments within these variables categories, which can be improved by a wider set of data.

The study results confirm that our proposed approach is suitable for recognizing the patterns of ROR crashes, when combining multiple large datasets at the state level, or even at the regional level. As such, the approach outlined in this study can highlight the contributions of EDA to all aspects of smart cities such as big data analytics and traffic safety risk of individual drivers. As for the total explained variances by the study variables, eigenvalue correction can be conducted on the Burt matrix to increase the variances (Das and Sun 2015, 2016; Abdi and Valentin 2007). Possible extension of this study can focus on, including person-level data to consider the effect of drivers' characteristics on crash occurrences. It should be noted that although the approach set forth here does not calculate the marginal effects of the variables, the ease of analyzing the big crash data following this approach in identifying the most statistically significant combinations of factors is exceptional. Moreover, taking advantage of the properties of MCA method, it is possible to not only identify contributing factors but also define associations between these factors. As such, MCA certainly has the potential to help state DOTs to select safety countermeasures that are associated with multiple safety benefits

REFERENCES

Abdi, H., and D. Valentin. Multiple Correspondence Analysis. *Encyclopedia of Measurement and Statistics*, 2007, pp. 651–657.

American Traffic Safety Services Association (ATSSA). *Preventing Vehicle Departures from Roadways*, 2015.

Baratian-Ghorghi, F., H. Zhou, M. Jalayer, and M. Pour-Rouholamin. Prediction of Potential Wrong-Way Entries at Exit Ramps of Signalized Partial Cloverleaf Interchanges. Traffic Injury Prevention, Vol. 16, No. 6, 2015, pp. 599–604.

Baratian-Ghorghi, F., H. Zhou, and I. Wasilefsky. Effect of Red-Light Cameras on Capacity of Signalized Intersections. Journal of Transportation Engineering, Vol. 142, No. 1, 2016.

Chatfield, C. *Problem Solving: A Statistician's Guide*. 2nd Edition, Chapman and Hall/CRC, United Kingdom, 1995.

Cook D., and D. F. Swayne. *Interactive and Dynamic Graphics for Data Analysis*. Springer-Verlag, New York, 2007.

Correspondence Analysis. *Numbers International Pty Ltd*. https://wiki.q-research-software.com/wiki/Correspondence_Analysis. Accessed May 1, 2016.

Das, S., and X. Sun. Factor Association Using Multiple Correspondence Analysis in Vehicle-Pedestrian Crashes. In *Transportation Research Record: Journal of Transportation Research Board*, No. 2519, Transportation Research Board of the National Academies, Washington, DC, 2015, pp. 95–103.

Das, S., and X. Sun. Association Knowledge for Fatal Run-off-Road Crashes by Multiple Correspondence Analysis. IATSS Research, Vol. 39, No. 2, 2016, pp. 146–155.

De Albuquerque, F. D. B., D. L. Sicking, and C. S. Stolle. Roadway Departure and Impact Conditions. In *Transportation Research Record: Journal of Transportation Research Board*, No. 2195, Transportation Research Board of the National Academies, Washington, DC, 2010, pp. 106–114.

Dissanayake, S., and U. Roy. Crash Severity Analysis of Single Vehicle Run-off-Road Crashes. Journal of Transportation Technologies, Vol. 4, No. 1, 2014, pp. 1–10.

Eustace, D., O. Almuntairi, P. W. Hovey, and G. Shoup. Using Decision Tree Modeling to Analyze Factors Contributing to Injury and Fatality of Run-off-Road Crashes in Ohio. *The 93rd Annual Meeting of Transportation Research Board*, Washington, DC, 2014.

Factor, R., G. Yair, and D. Mahalel. Who by Accident? The Social Morphology of Car Accident. Journal of Risk Analysis, Vol. 30, No. 9, 2010, pp. 411–1423.

Federal Highway Administration (FHWA). *Roadway Departure Safety*. www.safety.fhwa.dot.gov/roadway_dept. Accessed May 1, 2016. N.d.a

Federal Highway Administration (FHWA). *Safety Compass: Highway Safety Solutions for Saving Lives*. www.safety.fhwa.dot.gov/roadway_dept. Accessed May 1, 2016. N.d.b

Fitzpatrick, C. D. *The Effect of Roadside Elements on Driver Behavior and Run-off-the-Road Crash Severity*. M.Sc. Dissertation, University of Massachusetts Amherst, Amherst, Massachusetts, 2013.

Fontaine, H. A. Typological Analysis of Pedestrian Accidents. *Seventh ICTCT Workshop*, Paris, France, 1995.

Gates, T., P. Savolainen, T. Datta, and B. Russo. Exterior Roadside Noise Associated with Centerline Rumble Strips as a Function of Depth and Pavement Surface Type. Journal of Transportation Engineering, Vol. 140, No. 3, 2014.

Gifi, A. *Nonlinear Multivariate Analysis*. John Wiley & Sons, Hoboken, NJ, 1991.

Greenacre, M., and J. Blasius. *Multiple Correspondence Analysis and Related Methods*, 1st Edition, Chapman and Hall/CRC, United Kingdom, 2006.

Hallmark, S. L., Y. Qiu, M. Pawlovitch, and T. J. McDonald. Assessing the Safety Impacts of Paved Shoulders. Journal of Transportation Safety and Security, Vol. 5, No. 2, 2013, pp. 131–147.

Hallmark, S. L., S. Tyner, N. Oneyear, C. Carney, and D. McGehee. Evaluation of Driving Behavior on Rural 2-Lane Curves Using the SHRP 2 Naturalistic Driving Study Data. Journal of Safety Research, Vol. 54, 2015, pp. 17–27.

Heaslip, K., R. Bosworth, R. Barnes, A. Soltani-Sobh, M. Thomas, and Z. Song. Effects of Natural Gas Vehicles and Fuel Prices on Key Transportation Economic Metrics (No. WA-RD 829.1), Washington State Department of Transportation, Olympia, WA, 2014.

Jalayer, M., J. Gong, H. Zhou, and M. Grinter. Evaluation of Remote Sensing Technologies for Collecting Roadside Feature Data to Support Highway Safety Manual Implementation. Journal of Transportation Safety and Security, Vol. 7, No. 4, 2015, pp. 345–357.

Jalayer, M., and H. Zhou. A Sensitivity Analysis of Crash Prediction Models Input in the Highway Safety Manual. *The 2013 ITE Midwest District Conference*, Milwaukee, Wisconsin, 2013.

Jalayer, M., and H. Zhou. Evaluating the Safety Risk of Roadside Features for Rural Two-Lane Roads Using Reliability Analysis. Journal of Accident Analysis and Prevention, Vol. 93, 2016a, pp. 101–112.

Jalayer, M., and H. Zhou. Overview of Safety Countermeasures for Roadway Departure Crashes. ITE Journal, Vol. 86, No. 2, 2016b, pp. 39–46.

Khalilikhah, M., and K. Heaslip. Analysis of Factors Temporarily Impacting Traffic Sign Readability. International Journal of Transportation Science and Technology, 2016a.

Khalilikhah, M., and K. Heaslip. Important Environmental Factors Contributing to the Temporary Obstruction of the Sign Messages. In *Transportation Research Board 95th Annual Meeting* (No. 16-3785), 2016b.

Kim, K., and E. Yamashita. Corresponding Characteristics and Circumstances of Collision-Involved Pedestrians in Hawaii. In *Transportation Research Record: Journal of the Transportation Research Board*, No. 2073, Transportation Research Board of the National Academies, Washington, DC, 2008, pp. 18–24.

Lee, J., and F. L. Mannering. *Analysis of Roadside Accident Frequency and Severity and Roadside Safety Management*. Publication WA-RD 475.1. Washington State Department of Transportation (WSDOT), 1999.

Lee, J., and F. L. Mannering. Impact of Roadside Features on the Frequency and Severity of Run-off-Roadway Accidents: An Empirical Analysis. Journal of Accident Analysis and Prevention, Vol. 34, No. 2, 2002, pp. 149–161.

LeRoux, B., and H. Rouanet. *Multiple Correspondence Analysis*. SAGE Publications, Inc., United States, 2010.

Liu, C., and R. Subramanian. *Factor Related to Fatal Single-Vehicle Run-off-Road Crashes*. Publication DOT-HS-811-232. U.S. Department of Transportation, Washington, DC, 2009.

Liu, C., and T. J. Ye. *Run-off-Road Crashes: An On-Scene Perspective*. Publication DOT-HS-811-500. U.S. Department of Transportation,, Washington, DC, 2011.

McGinnis, R. G., M. J. Davis, and E. A. Hathaway. Longitudinal Analysis of Fatal Run-off-Road Crashes, 1975 to 1997. In *Transportation Research Record: Journal of Transportation Research Board*, No. 1746, Transportation Research Board of the National Academies, Washington, DC, 2001, pp. 47–58.

McLaughlin, S. B., J. M. Hankey, S. K. Klauer, and T. A. Dingus. *Contributing Factors to Run-off-Road Crashes and Near-Crashes*. Publication DOT-HS-811-079. National Highway Traffic Safety Administration (NHTSA), Washington, DC, 2009.

Montella, A. Safety Evaluation of Curve Delineation Improvements. In *Transportation Research Record: Journal of Transportation Research Board,* No. 2103, Transportation Research Board of the National Academies, Washington, DC, 2009, pp. 69–79.

Nallet, N., M. Bernard, and M. Chiron. Individuals Taking a French Driving Licence Points Recovery Course: Their Attitudes Towards Violations. Accident Analysis & Prevention, Vol. 40, No. 6, 2008, pp. 1836–1843.

National Highway Traffic Safety Administration (NHTSA). *Fatality Analysis Reporting System (FARS).* www.nhtsa.gov/FARS. Accessed May 1, 2016.

Neuman, T. R., R. Pfefer, K. L. Slack, K. K. Hardy, F. Council, H. McGee, L. Prothe, and K. Eccles. *A Guide for Addressing Head-on Collisions.* National Cooperative Highway Research Program (NCHRP), Report No. 500(4), Washington, DC, 2003a.

Neuman, T. R., R. Pfefer, K. L. Slack, K. K. Hardy, F. Council, H. McGee, L. Prothe, and K. Eccles. *A Guide for Addressing Run-off-Road Collisions.* National Cooperative Highway Research Program (NCHRP), Report No. 500(6), Washington, DC, 2003b.

Office of the Assistant Secretary for Research and Technology (OST-R). *usRAP: A New Tool for Road Safety Management.* www.rita.dot.gov/utc/publications/spotlight/2009_11/html/spotlight_0911.html. Accessed May 1, 2016.

Panagiotakos, D. B., and C. Pitsavos. Interpreting of Epidemiological Data Using Multiple Correspondence Analysis and Log-Liner Models. Journal of Data Science, Vol. 2, 2004, pp. 75–86.

Peng, Y., and L. N. Boyle. Commercial Driver Factors in Run-off-Road Crashes. In *Transportation Research Record: Journal of Transportation Research Board,* No. 2281, Transportation Research Board of the National Academies, Washington, DC, 2012, pp. 128–132.

Petegem, J. W. H., and F. Wegman. Analyzing Road Design Risk Factors for Run-Off-Road Crashes in the Netherlands with Crash Prediction Models. Journal of Safety Research, Vol. 49, 2014, pp. 121–127.

Pour-Rouholamin, M., and M. Jalayer. Analyzing the Severity of Motorcycle Crashes in North Carolina Using Highway Safety Information Systems Data. ITE Journal, Vol. 86, No. 10, 2016, pp. 45–49.

Roque, C., and J. L. Cardoso. Safeside: A Computer-Aided Procedure for Integrating Benefits and Costs in Roadside Safety Intervention Decision Making. Journal of Safety Science, Vol. 74, 2015, pp. 195–205.

Roque, C., F. Moura, and J. L. Cardoso. Detecting Unforgiving Roadside Contributors Through the Severity Analysis of Ran-off-Road Crashes. Journal of Accident Analysis and Prevention, Vol. 80, 2015, pp. 262–273.

Satterfield, C. *Federal Highway Administration Focus Area Data Definitions.* Publication No. FHWA-HRT-14-062. 2014.

Savolainen, P. T, F. L. Mannering, D. Lord, and M. A. Quddus. The Statistical Analysis of Highway Crash-Injury Severities: A Review and Assessment of Methodological Alternatives. Journal of Accident Analysis and Prevention, Vol. 43, 2011, pp. 1666–1676.

Schrum, K. D., F. D. B. D. Albuquerque, D. L. Sicking, R. K. Faller, and J. D. Reid. Benefits of Slope Flattering. Journal of Transportation Safety and Security, Vol. 4, 2014, pp. 356–368.

Sharifi, M. S., and H. Shabaniverki. Modeling Crash Delays in a Route Choice Behavior Model for Tow Way Road Network. Journal of Geotechnical and Transportation Engineering, Vol. 2, No. 1, 2016.

Sharifi, M. S., D. Stuart, K. Christensen, A. Chen, Y. S. Kim, and Y. Chen. Analysis of Walking Speeds Involving Individuals with Disabilities in Different Indoor Walking Environments. Journal of Urban Planning and Development, Vol. 142, No. 1, 2015.

Soltani-Sobh, A., K. Heaslip, R. Bosworth, and R. Barnes. Effect of Improving Vehicle Fuel Efficiency on Fuel Tax Revenue and Greenhouse Gas Emissions. In *Transportation Research Record: Journal of the Transportation Research Board,* No. 2502, Transportation Research Board of the National Academies, Washington, DC, 2015, pp. 71–79.

Spainhour, L. K., and A. Mishra. Analysis of Fatal Run-off-the-Road Crashes Involving Overcorrection. In *Transportation Research Record: Journal of Transportation Research Board,* No. 2069, Transportation Research Board of the National Academies, Washington, DC, 2008, pp. 1–8.

United States Road Assessment Program (usRAP). *usRAP Coding Manual for Star Ratings and Safer Roads Investment Plans* Roadway Safety Foundation, Chicago, Illinois, 2012.

United States Road Assessment Program (usRAP). *usRAP Study of Eight Counties in Illinois.* CH2M Hill, Chicago, Illinois, 2013.

Wu, K., E. T. Donnell, and J. Aguero-Valverde. Relating Crash Frequency and Severity: Evaluating the Effectiveness of Shoulder Rumble Strips on Reducing Fatal and Major Injury Crashes. Journal of Accident Analysis and Prevention, Vol. 67, 2004, pp. 86–95.

Ye, F., and D. Lord. Comparing Three Commonly Used Crash Severity Models on Sample Size Requirements: Multinomial Logit, Ordered Probit and Mixed Logit Models. Journal of Analytic Methods in Accident Research, Vol. 1, 2014, pp. 72–85.

Zhou, H., M. Jalayer, J. Gong, S. Hu, and M. Grinter. *Investigation of Methods and Approaches for Collecting and Recording Highway Inventory Data.* Publication FHWA-ICT-13-022. Illinois Center for Transportation, 2013.

9 Predicting Traffic Safety Risk Factors Using an Ensemble Classifier

Nasim Arbabzadeh, Mohammad Jalayer, and Mohsen Jafari

CONTENTS

9.1 INTRODUCTION

Motor vehicle traffic crashes are one of the leading causes of accidental death, killing over 35,000 people in the United States in 2015 alone (NHTSA 2015). The United States Department of Transportation (USDOT) has observed that 2015 marked the highest increase in traffic deaths for 49 years (NHTSA 2016). The USDOT has also indicated that estimates of 2016 fatalities show another frightening uptick of more than 10 percent. Beyond the loss of human life, these accidents also cause over $870 billion in costs each year (NHTSA n.d). While there is no singular cause of vehicle crashes and a combination of human, roadway, and vehicle factors contribute in crash occurrences, human errors (e.g., impaired conditions, inadvertent errors, and risky driving behavior) are associated with nearly 90 percent of light-vehicle crashes. Driver error is also the primary contributing factor for approximately 87 percent of all commercial vehicles crashes (FMCSA 2006).

Given this fact, to enhance the overall safety in road networks, safety management approaches must focus more on individual driver's behavior. We note that there has been a considerable number of studies identifying the significant factors contributing to motor vehicle crashes, using historical crashes (Das and Sun 2016; Factor et al. 2010; Nallet et al. 2008; Kim and Yamashita 2008; Fontaine 1995; Lee and Mannering 2002; McGinnis et al. 2001; Wu et al. 2014; Jalayer and Zhou 2017). These studies attempted to link the likelihood of crash occurrences at different severity levels to the characteristics of roadways, drivers, and environments to mitigate the crash risk in crash-prone locations. This reactive approach only uses the police-reportable crashes while the supporting crash data is often dated, incomplete and insufficient to support accurate diagnosis and intervention. With the emerging trends in smart transportation and infrastructure, the widespread use of advanced technology (e.g., sensors, radars, cameras, and onboard vehicular devices), and advances in big data storage and analytics, recording and processing of high-quality traffic safety data will be readily available. To create a high-resolution insight into the driving events, it is possible to obtain the safety-related traffic data from these sources and overlaid over time and location for a particular driver. Furthermore, these driving data can be linked with supplementary roadway and weather data to build a holistic view of events and consequential behavioral patterns and safety factors. Several studies (Xu et al. 2013; Hossain and Muromachi 2012; Shew et al. 2013; Xu et al. 2014; Ahmed and Abdel-Aty 2013; Yu and Abdel-Aty 2013) developed a prediction of real-time risk of crashes using loop detector data. These works estimate the likelihood of crash occurrence for a given freeway segment over a short period without considering the driver's personalized safety factors contributing to the crash.

This paper sought to fill the gap in knowledge by defining and establishing the real-time individualized traffic safety risk as the likelihood of a crash and near-crash events and model its relationship to safety factors using a data-driven and learning-based algorithm. We used real-world 100-car naturalistic driving studies (NDS) dataset, namely VTTI's 100-car, to develop an ensemble of Breiman's random forest and a newly proposed Multivariate Time Series Random Forest to classify driving events into the crash and near-crash classes on a set of safety factors. The replicated *k*-fold cross validation is used to evaluate the models' performances. The results of this study will provide useful insight into human factors contributing to crash and near-crash events. This paper also provides car insurance company valuable information for the application of the behavior-based auto insurance. Moreover, this study assists researchers, engineers, policymakers, and transportation agencies to obtain a greater understating of errors and human-related contributing factors in crashes with the aim of developing effective strategies to mitigate the crash-injury severity outcomes.

9.2 PROBLEM STATEMENT

We denote the driving outcome or state for driver *i*, on his/her trip *j*, at location *l*, and at time *t* by Y_{ijlt}, therefore making it driver-specific. Assuming that driving state follows a categorical distribution, in this paper, we split the driving outcome's spectrum into three different discrete categories including normal-driving, near-crash, and crash events as follows:

$$Y_{ijlt} = \begin{cases} 1: & \text{The state is normal driving} \\ 2: & \text{The state is a near-crash} \\ 3: & \text{The state is a crash} \end{cases} \tag{9.1}$$

It should be noted that a categorical distribution is a probability distribution that defines all possible outcomes of a random phenomenon that can undertake one of the several possible outcomes, in which the probability of each occurrence separately specified. These probabilities are constrained only by the fact that each probability must be in the range zero to one, and they all must sum to one. Figure 9.1 illustrates the state of driving as a near-continuum spectrum of colors, where each driving state is related to several sub-states or colors. The spectrum ranges from the least (a near-zero chance of conflict) to the highest (a severe crash) level of driving safety risk.

There can be different approaches to mapping the driving state to the safety risk. In this study, we defined the safety risk as the probability of occurrence of

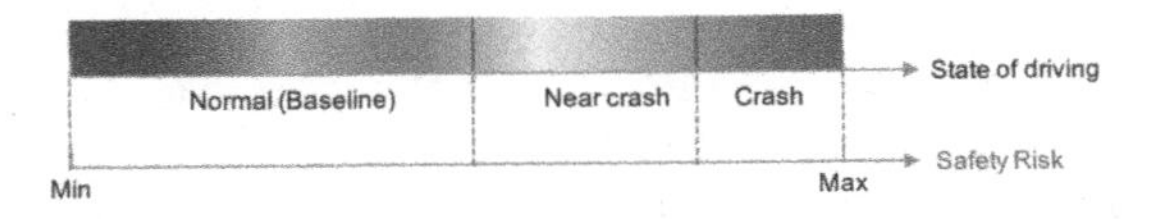

Figure 9.1 Spectral driving outcome.

critical events (i.e., crash and near-crash events). We denote and define the vector of independent safety factors of length p by Z_{ijlt} as follows:

$$Z_{ijlt} = \left(xD_{ij\circ\circ}, xV_{ij\circ\circ}, x_{ijlt}, u_{ijlt}, xC_{ijlt}\right) \tag{9.2}$$

where xD_{ijoo} and xV_{ijoo} are the static safety factors associated with the characteristics of the target driver and vehicle, respectively. Note that in this paper, the term "static" is referred to both the invariant factors such as driver's gender and vehicle's make/model and the (low frequency) changing factors such as vehicle's maintenance condition and driver's driving experience. The latter typically do not change significantly during an individual trip but may be different from one trip to another or during longer time intervals (month, season or year). Vectors x_{ijlt} and u_{ijlt} include information about the vehicle kinematics and control variables, respectively, similar to the state and input vectors in the state-space representation of physical systems. Vector xC_{ijlt} encompasses all other variables describing the context of driving (e.g., driver's dynamic behavior, time of driving, roadway design features, and weather condition). Given a pair of datasets (Z_{ijlt}, Y_{ijlt}), we aim to develop a function, namely f, which can best model the relationship between the safety factors, Z_{ijlt}, and the driving state, Y_{ijlt}:

$$f : Z_{ijlt} \rightarrow Y_{ijlt} \tag{9.3}$$

For simplicity purposes and to consider storage capacity, the continuous time t is substituted by kT_s, where k is greater than zero and T_s is a constant time step. Despite the fact several factors such as vehicle's kinematic variables (e.g., speed and acceleration) are high-frequency time-varying variables, other factors such as number of travel lanes changes less frequently, assuming to be constant during a small time horizon, NT_S, where N is the number of time steps in the horizon or length of the time-series. Given these facts, we divided the safety variables into *time-series* ($T_{ij(l,L),(K,N)}$) and *event* ($ev_{ij(l,L),(K,N)}$) variables based on the frequency of real-time changes and denoted them as follows:

$$T_{ij(l,L),(k,N)} = \left(\begin{pmatrix} \vec{x}ij, l-L, k-N \\ \vec{u}ij, l-L, k-N \end{pmatrix}'_{1\times p_1}, \ldots, \begin{pmatrix} \vec{x}ijlt \\ \vec{u}ijlt \end{pmatrix}'_{1\times p_1} \right)_{N\times p_1} \tag{9.4}$$

$$ev_{ij(l,L),(k,N)} = xC_{ij(l,L),(k,N)1\times p_2} \tag{9.5}$$

Considering the above discrete-time representation and the new categorization of safety factors into the *time-series* and *event* variables, we redefined the vector of safety predictors, Z_{ijlt}, to an array of vector and scalar, $Z_{ij(l,L),(K,N)}$, as follows:

$$z_{ij(l,L),(k,N)} = \left(xD_{ij\circ\circ}, xV_{ij\circ\circ}, \begin{pmatrix} x_{ij,l-L,k-N} \\ \vdots \\ x_{ijlt} \end{pmatrix}, \dots, \begin{pmatrix} u_{ij,l-L,k-N} \\ \vdots \\ u_{ijlt} \end{pmatrix}, xC_{ij(l,L),(k,N)} \right) \tag{9.6}$$

$$= \left(xD_{ij\circ\circ}, xV_{ij\circ\circ}, T_{ij(l,L),(k,N)}, ev_{ij(l,L),(k,N)} \right)$$

As mentioned earlier, we define the traffic safety risk as the likelihood of an unfavorable outcome. The safety risk is computed based on the following conditional probability:

$$P_a = \Pr\left(y_{ijkl} = a \mid Z_{ij(l,L),(k,N)}\right) = f\left(Z_{ij(l,L),(k,N)}\right); \; for\ a \in S_{Cr} \tag{9.7}$$

where S_{Cr} is the set of critical outcomes and f is a data-driven function. The predicted state of driving can be computed as follows:

$$\hat{y}_{ijkl} = \underset{a=1,2,\dots,C}{Arg\max}\, P_a \tag{9.8}$$

9.3 METHOD

To calculate the predicted class of the driving outcome, we employed the ensemble classifier, which is mainly used to enhance the performance of a model for classification, prediction, and function approximation (Polikar 2008). An ensemble-based system is acquired by merging different models, hereafter classifiers. Some situations make use of an ensemble-based system statistical sense. For example, in many applications of automated decision-making, it is quite common to obtain data from various sources, providing complementary information. An appropriate combination of such information is typically referred to as data or information fusion. Compared to a stand-alone decision support by any of the individual data sources, the data fusion increases the accuracy of the classification decision.

Sometimes, heterogeneous features, such as *time-series* and *event* variables in Equations 9.4 and 9.5, cannot be employed all together to train a single classifier and even if they could such training may be more unlikely to be successful. In such cases, an ensemble of classifiers will be a good selection where a separate classifier is trained on each of the feature sets independently of others. A combination rule can then integrate the decisions made by each classifier. We note that there are two different types of combining rules: algebraic (or fixed) rules and trained rules. The algebraic rules combine the continuous-valued output of classifiers through an algebraic expression such as sum, mean, minimum, maximum, and median. Following the algebraic expression is applied to individual supports acquired by each class, the final ensemble decision in each event is a class that gathers the major support. As per our earlier notations, P_{ia} is the conditional probability of output j, obtained by applying classifier M_i, and denotes as follows:

$$P_{ia} = \Pr\left\{Y_{ijkl} = a \mid T_{ij(l,L),(k,N)}, ev_{ij(l,L),(k,N)}, M_i\right\}; \quad a \in S_{Cr} \text{ and } i=1,2,\dots,m \tag{9.9}$$

where:

M_i is classifier i

m is the number of classifiers

Therefore, P_a in Equation 9.7 can be calculated as follows:

$$P_a = \Pr\left\{Y_{ijkl} = a \mid T_{ij(l,L),(k,N)}, ev_{ij(l,L),(k,N)}\right\} = \frac{\text{rule}_i P_{ia}}{\sum_c \text{rule}_i P_{ic}} \quad (9.10)$$

where $\text{rule}_i P_{ia}$ provides the combined value of continuous-valued output a overall classifiers. The class of event scenario, $\{T_{ij(l,L),(k,N)}, ev_{ij(l,L),(k,N)}\}$, can then be obtained from Equation 9.8. It should be noted that one can train an arbitrary classifier using the values of P_{ia} (for all i and a) as features in the intermediate space; therefore, the combining rule is called a trained rule. For the purpose of this paper, we used fixed combing rules.

As discussed earlier, there are two different types of safety variables based on the frequency of real-time changes including *time-series* variables and *event* variables. In a recent study, Jafari et al. (2014) applied multinomial logistic regression (MLR) model on *event* variables of VTTI's 100-car NDS data to classify the driving scenarios. In this paper, we used Breiman's random forest (RF) method to classify *event* variables. We conducted a replicated cross validation to evaluate and compare the performances of both MLR and RF methods on the classification of *event* data. To classify the *time-series* variables, we employed a recent generalization of RFs for multivariate time series by Baydogan and Runger (2015), as discussed later.

9.4 CLASSIFICATION OF EVENT DATA

9.4.1 Multinomial Logistic Regression (MLR)

To estimate the trichotomous driving outcome (i.e., normal-driving (baseline), near-crash, and crash events), we used the MLR model. Logistic regression technique is designed to estimate the parameters of a multivariate explanatory model where the dependent or response variable is dichotomous, and the independent variables are continuous or categorical (Dutta et al. 2015). MLR is an extension of logistic regression to multiple outcome categories (Gelman and Hill 2007). The predicted values from the analysis can be interpreted as the probability of membership to the target groups. MLR uses a linear predictor function to predict the likelihood that observation i takes the outcome k, as shown in Equation 9.11.

$$\text{logit} \Pr\left\{Y_{ijlt} = a \mid ev_{ij(l,L),(k,N)}\right\} = \beta_0{}^a + \beta^a . ev_{ij(l,L),(k,N)} \quad (9.11)$$

where $(\beta_0{}^a, \beta^a)$ is the vector of regression coefficients associated with outcome a. For C possible outcomes, C-1 independent binary logistic regression models are built, so that one outcome is selected as a pivot or reference category and the other C-1 outcomes are separately regressed against the reference category.

9.4.2 Random Forest (RF)

Random forest is an ensemble classifier that comprises of many decision trees and outputs the majority vote of individual trees (Hastie et al. 2009). Regression trees assume a model of the form:

$$f(x) = \sum_{m=1}^{M} c_m . 1(X \in R_m) \quad (9.12)$$

where $R_1,\ldots,R_M$ represent a partition of feature space. We note that trees are constants under scaling and various other transformations of feature values. Moreover, trees are robust to inclusion of dissimilar features using inspectable production rules. However, they incline to learn extremely irregular patterns and overfit the training sets, all of which lead to low bias and high variance. In order to decrease the variance, RF averages multiple deep decision trees and trains on various parts of the same training set. This results in a small increase in the bias and the loss of interpretability. However, it significantly boosts the performance of the final model. RF creates numerous classification trees, and each tree gives a vote at the input vector by providing a classification (Suma et al. 2014; Kamei and Shihab 2016). Considering the input vector on each tree in the forest, RF classifies a new object from an input vector. The forest takes the classification with the most votes over all trees in the forest (Suma et al. 2014; Kamei and Shihab 2016).

9.4.3 Method Selection

To evaluate the prediction performance of MLR and RF classifiers on *event* data, we used *k*-fold cross validation (CV), which is an extensively used technique for verifying the robustness of a model (Yang et al 2014). The original sample in the *k*-fold CV is randomly divided into *k* equal parts or subsets. Of the *k* subsets, a subset, as a means to validate data, is taken to test the model and the remaining subsets, *k*-1, are used as training data. This process is repeated *k* times or folds in which each subset is engaged once as the validation data. Thereupon, the results of the *k* folds are averaged to create a single estimation. It should be noted that *k* is an unfixed parameter, and its best value can be identified through experiments. The *k*-fold CV misclassification error rate takes the form:

$$\begin{aligned} kCV\_MCER &= \frac{1}{k}\sum_{i=1}^{k} MCER_i \\ &= \frac{1}{k}\sum_{i=1}^{k}\frac{1}{N_i}\sum_{j\in Fold\,i} I(y_j \neq y_j) \end{aligned} \tag{9.13}$$

where *MCER* is the mean CV classification error, *k* is the number of subsets, and N_i is the number of instances in fold *i*. Given the low frequency of crash and near-crash events, compared to the baseline (normal-driving) events, the traffic safety data is highly imbalanced. It means that the classification categories are not represented approximately equally. To be specific, in 100-car NDS data, the proportions of classification categories for the crash, near-crash, and baseline events are 68, 760, and 19,000, respectively. To overcome this issue, we employed stratified *k*-fold CV in which the folds are chosen to make the mean response value in all folds (approximately) equally. In this study, we performed both *k*-fold CV and stratified *k*-fold CV techniques to evaluate the performances of MLR and RF models using the *event* data.

9.4.4 Multivariate Time Series Random Forest (MTS-RF)

In recent decades, multivariate time series (MTS) classification has received special attention in the fields of medicine, finance, and multimedia due to increase in the number of temporal datasets. Baydogan and Runger (2015) present a classifier based on a new exemplification for MTS and denote as multivariate time series classification (SMTS). Compared to the MTS method that considers all attributes separately, SMTS considers the attributes of MTS simultaneously to capture the information associated with relationships between classes. Considering this fact, an equivalent formulation of our time-series classification problem is as follows:

$T_{ij(l,L),(K,N)}$ is a p_1-attribute time series each of which has N observations where t_m^n is the m-th attribute or safety factor of series n and $t_m^n(k)$ denotes the observation at time step k. Time series can be of different sizes and MTS-RF handles this situation, but for the purpose of illustration here we assume time series to be of the same length. MTS example T^n is represented by a $N \times p_1$ matrix as follows:

$$T^n = \left[t_1^n, t_1^n, \ldots, t_m^n, \ldots, t_{p1}^n\right] \tag{9.14}$$

$$t_m^n = \left[t_m^n(k-N), \ldots, t_m^n(k)\right] \tag{9.15}$$

where t_m^n is the time series in column m. There are N' training MTS, each of which is associated with a class label $Y_n \in \{1,2,\ldots,C\}$ for $n = 1, 2, \ldots, N'$. Given a set of unlabeled MTS, the task is to map each MTS to one of the predefined classes. Rather than extracting features from each time series, each row of T^n is considered to be an instance. This is achieved by creating a matrix of instances $D_{N'N \times p_1}$:

$$D_{N'N \times p_1} = \begin{bmatrix} t_1^1 & t_2^1 & \cdots & t_{p1}^1 \\ t_1^2 & t_2^2 & & t_{p1}^2 \\ \vdots & & \ddots & \vdots \\ t_1^{N'} & t_2^{N'} & \cdots & t_{p1}^{N'} \end{bmatrix} \tag{9.16}$$

Equation 9.14 is the concatenation of training examples T^n in Equation 9.12. We assign the label of each instance to be the same as the time series. Then, $D_{NN \times p_1}$ is mapped to the feature space $\Phi_{N'N \times (2p_1+1)}$ that adds the following columns: time index, first differences for each numerical attribute. The row of Φ for series n at time step k is:

$$\left[k, t_1^n(k), t_1^n(k) - t_1^n(k-1), \ldots, t_{p1}^n(k), t_{p1}^n(k) - t_{p1}^n(k-1)\right] \tag{9.17}$$

The differences provide trend information. A tree learner can capture this information if it relates to the class. Given the fact that time is used as a predictor variable and RFs can efficiently handle interactions, complex regions in two-dimensional signal space (S), where one class dominates, can be detected. In this sense, the time ordering of the data is used. Following the generation of the symbolic representation of the trees in RFins, a bag-of-words approach is used to classify the time series. The readers are referred to Baydogan and Runger (2015) for more detailed information about this method.

9.5 DATA

To examine the performance of our method, we used 100-car NDS data which includes information on drivers' driving behavior and performance (e.g., fatigue, impaired driving, judgment error, distracted driving, aggressive driving, and traffic violations) as event occurrence in real-time. The 100-car NDS is the first instrumented-vehicle study conducted to gather large-scale, naturalistic driving data in the United States and includes more than 1 year of collected data, resulting in about 2,000,000 vehicle miles of driving and nearly 43,000 hours of driving data (NHTSA 2006). Following data collection, two different databases were created including *events* and *baseline*. The former includes crash and near-crash events while the latter contains regular driving incidents. A crash is defined as any physical contact between the subject vehicle and another vehicle, pedestrian, pedacyclist, fixed object, animal, etc. A near-crash is a situation in which a rapid, severe evasive maneuver needed to avoid a crash (NHTSA 2006). We note that the number of epochs selected per vehicle in the baseline database is proportional to the number of vehicles contributed to crashes, near-crashes, and incidents.

NDS database includes two major types of data: *time-series* data and *event* or *video-reduced* data. *Time-series* data contains direct readings from onboard devices (e.g., radars, sensors, and accelerometers) and is available for 68 crashes and 760 near-crash events. For each driving event, the dataset contains time-series variables such as gas pedal position and speed vehicle composite, spanning 30 seconds before and 10 seconds after an event. *Video-reduced* data includes detailed event, driver state, and driving environment information derived from video reduction for the same crash and near-crash events. Given the fact that time-series variables are not yet available for baseline events, we present the numerical results of our general model for the dichotomous problem of crash and near-crashes. Based on the recorded vehicle kinematics measured by accelerometers and gyroscope sensors, crash and near-crash events are usually distinguished. Moreover, by applying an "event" button designed in vehicles, the driver had an opportunity to record a driving event. To determine the additional events, vehicle kinematics along with rear and forward time-to-collision were also used. Following this process, the camera videos were reviewed to verify the safety-related events. Once the events confirmed, several parameters were recorded for each event including pre-event maneuver, precipitating factors contributing factor, associative factor, and avoidance maneuver.

Table 9.1 presents the list of the 25 variables ($p = 25$) included in our model. Of the 25 variables, 15 factors are *event* variables ($p_1 = 15$) including five driver-related, two environmental conditions, six roadway-characteristics, and two surrounding-externalities variables. The remained variables in the model, time-series variables ($p_2 = 10$), all are driver-related safety factors except the *Lighting*, explaining an externality. This table also illustrates the source and type of each variable. The readers are referred to (http://www.vtti.vt.edu/) for further explanation about the database and variables descriptions.

9.6 RESULTS AND DISCUSSIONS

To evaluate the performance of our approach, we used VTTI's naturalistic driver behavior data. Following our proposed methodology, we first used CV to compare the prediction performances of MLR and RF classifiers on *event* data. Figure 9.2 depicts the results of k-fold CV and stratified k-fold CV for $k = 2, 3, \ldots, 10$. We also performed replicated CV with $n = 10$ to smooth the error-rate values over k.

According to this figure, the RF classifier outperforms the multinomial logistic regression (MLR) model on feature space of the dichotomous *event* data. It also has a pretty robust performance excelling MLR model over all values of k with an error rate close to 8 percent. Therefore, we selected the RF classifier over the MLR model to calculate the driver's risk, i.e. the probabilities P_{ia}'s in Equation 9.8, and set $i = 1$ for the RF classifier given only the *event* data.

Following this, we employed the MTS-RF to classify the dichotomous output of crash and near-crash events on the time-series feature space. To evaluate the prediction performance of our classification model for a new unobserved data point, we ran a 5-fold CV. The best values in terms of out-of-bag error rate for the number of trees (150) and number of nodes (Polikar 2008) were used to grow the

Table 9.1: Input Variables to the Traffic Safety Risk Model

Variable Name	Time Dependency	Group	Source	Variable Type
Distraction	Snapshot	Driver	Internal	Categorical
Driver behavior	Snapshot	Driver	Internal	Categorical
Driver seatbelt use	Snapshot	Driver	Internal	Binomial
Subject age	Snapshot	Driver	Internal	Categorical
Subject gender	Snapshot	Driver	Internal	Binomial
Lighting	Snapshot	Environmental conditions	External	Categorical
Weather	Snapshot	Environmental conditions	External	Categorical
Alignment	Snapshot	Roadway-characteristics	External	Categorical
Locality	Snapshot	Roadway-characteristics	External	Categorical
Relation to junction	Snapshot	Roadway-characteristics	External	Categorical
Surface conditions	Snapshot	Roadway-characteristics	External	Categorical
Traffic control	Snapshot	Roadway-characteristics	External	Categorical
Travel lanes	Snapshot	Roadway-characteristics	External	Integer
Traffic density	Snapshot	Surrounding externalities	External	Categorical
Traffic flow	Snapshot	Surrounding externalities	External	Categorical
Gas pedal position	Time Series	Driver	Internal	Continuous
Speed vehicle composite	Time Series	Driver	Internal	Continuous
Speed GPS horizontal	Time Series	Driver	Internal	Continuous
Yaw rate	Time Series	Driver	Internal	Continuous
Heading GPS	Time Series	Driver	Internal	Continuous
Lateral acceleration	Time Series	Driver	Internal	Continuous
Longitudinal acceleration	Time Series	Driver	Internal	Continuous
Brake on/off	Time Series	Driver	Internal	Binomial
Turn signal state	Time Series	Driver	Internal	Categorical
Lighting	Time Series	Environmental conditions	External	Continuous

forest. In order to deal with the imbalanced data, we optimized the binary classification threshold with respect to sensitivity, specificity and positive predictive values. Figure 9.3 shows the Receiver Operating Characteristic curve and Table 9.2 shows the values of the abovementioned measures of goodness and the overall accuracy for different values of thresholds. It can be seen that if we set the binary classification threshold at 0.35, it will result in an overall accuracy of 98.2 percent, sensitivity (i.e., the probability of detection of crashes) of 79.4 percent while the precision rate (Positive Predictive Value, PPV) and specificity are 98 percent and 99 percent, respectively. Table 9.3 shows the confusion matrix of this classifier.

In the next step, we combined the results of RF classifier on *event* feature space (P_{1j}'s) with the results of MTS-RF on time-series feature space (P_{2j}'s). To that end, we designed stratified k folds of cases to train RF and MTS-RF separately on k-1 folds and calculate P_{1j}'s and P_{2j}'s on the left-out k-th fold. Then, by applying one of the fixed combining rules of ensemble classifiers introduced earlier, we calculated the final P_j's, i.e. the driving safety risks.

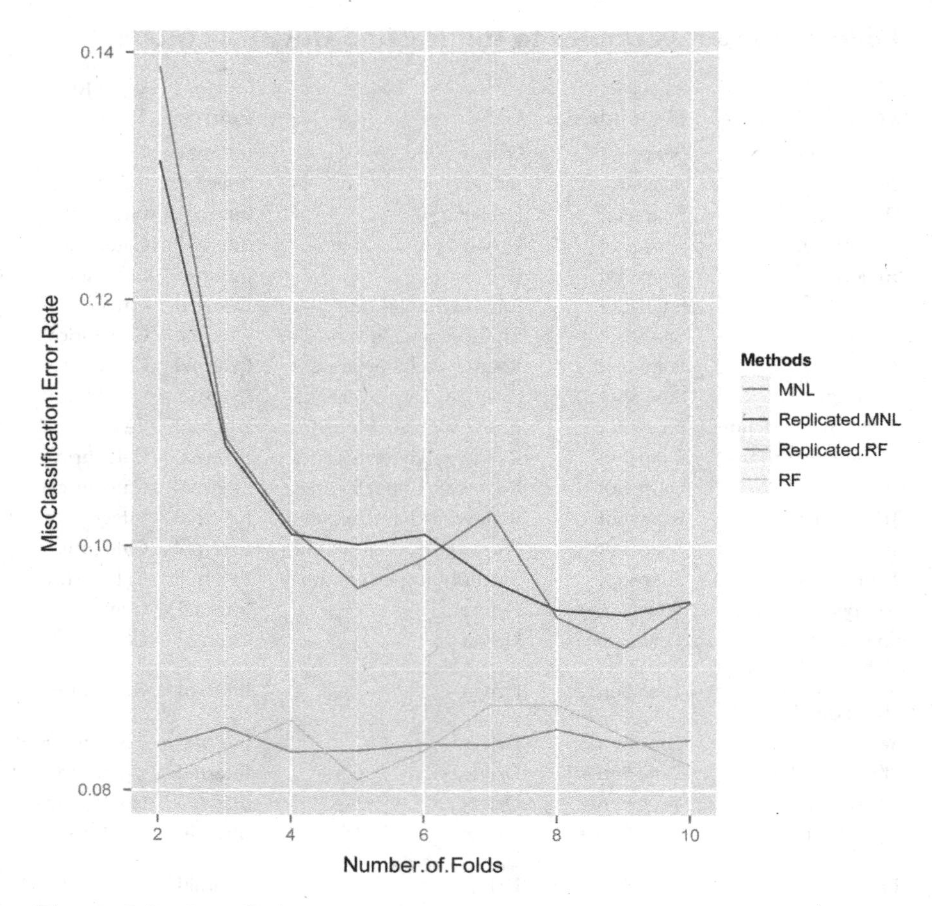

Figure 9.2 Stratified cross validation for the crash and near-crash events.

Figure 9.4 illustrates the results of stratified k-fold CV using RF on *event* data and MTS-RF on *time-series* data and ensemble classifiers with combining rules of minimum, maximum, mean, and product for the final decision fusion. For illustration purposes, results for $k = 5$ were illustrated since the same interpretations were concluded from all values of $k = 2,\ldots,10$. According to these figures, MTS-RF outperforms the other method, considering the lowest error rate.

In each trip and for each time step, we feed the state vector into the classification model to detect a crash and near-crash events for an individual driver. For convenience, we have color-coded the three safety states as follows:

9.7 CONCLUSIONS AND RECOMMENDATIONS

This paper developed a novel approach to formulate the real-time traffic safety risk of individual drivers and to predict the drivers' individualized safety risks (Figure 9.5). To evaluate the proposed frameworks, 100-car NDS dataset were employed. Taking advantages of an ensemble of Breiman's RF and a MTS-RF, we classified driving events into the crash and near-crash classes. To evaluate the models' performances, we used the replicated k-fold CV technique. To achieve

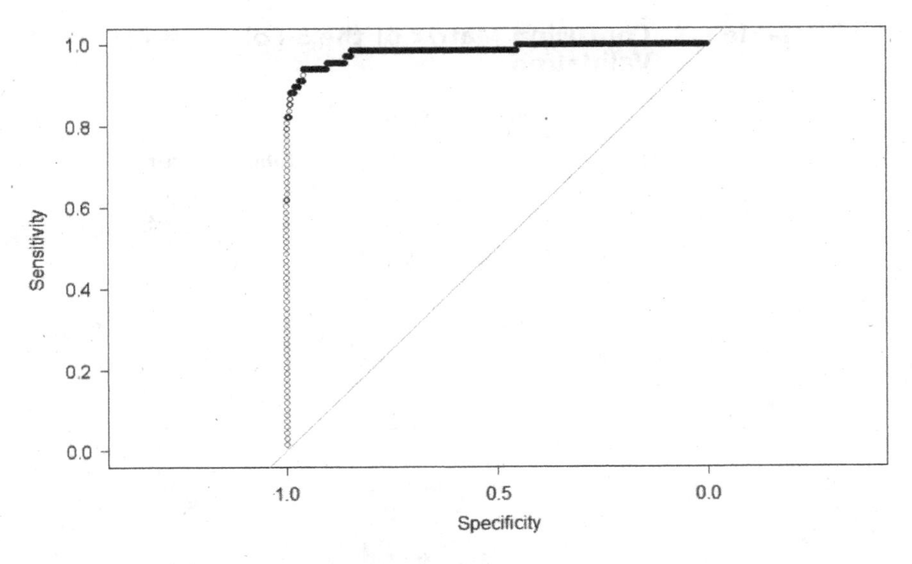

Figure 9.3 The result of specificity vs. sensitivity.

Table 9.2: Values of the Four Classification's Measures of Goodness for the MTS-RF

Threshold	Accuracy	Sensitivity	Specificity	Precision
0.00	0.082	1	0	0.082
0.05	0.844	0.985	0.83	0.344
0.10	0.925	0.941	0.92	0.525
0.15	0.964	0.912	0.97	0.721
0.20	0.975	0.882	0.98	0.822
0.25	0.979	0.853	0.99	0.892
0.30	0.982	0.824	1	0.949
0.35	0.982	0.809	1	0.965
0.40	0.979	0.765	1	0.981
0.45	0.973	0.691	1	0.979
0.50	0.969	0.632	1	0.977
0.55	0.967	0.603	1	1
0.60	0.965	0.574	1	1
0.65	0.963	0.544	1	1
0.70	0.963	0.544	1	1
0.75	0.961	0.529	1	1
0.80	0.961	0.529	1	1
0.85	0.958	0.485	1	1
0.90	0.954	0.441	1	1
0.95	0.937	0.235	1	1
1.00	0.919	0.015	1	1

Table 9.3: Confusion Matrix of the 5-Fold Cross Validation

		Predicted Classes			
		Crash	Near-Crash	Total	Accuracy
Actual Classes	Crash	55	13	68	80.9%
	Near-Crash	2	758	760	99.7%
	Precision	96%	98%		

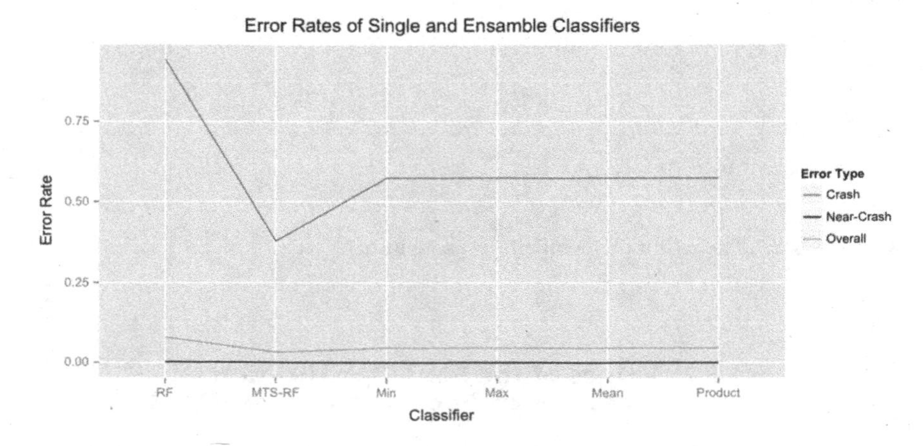

Figure 9.4 Average CV error rates of single and ensemble classifiers for different error types, N folds = 5.

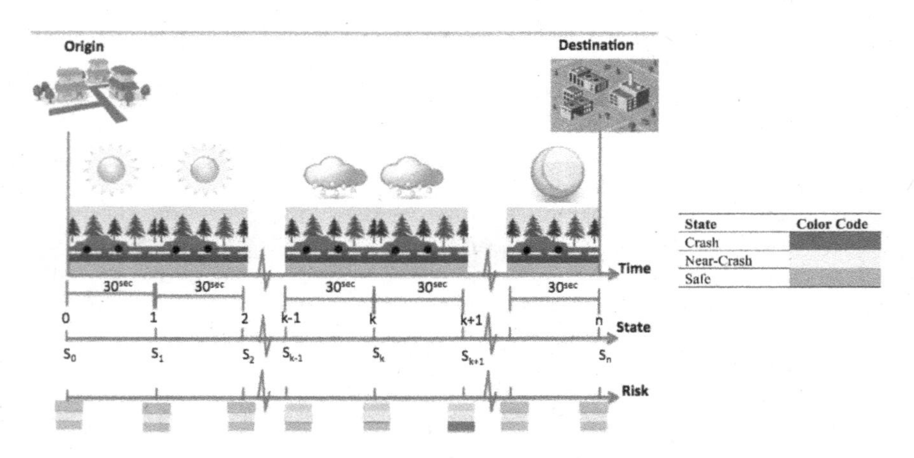

Figure 9.5 Schematic illustration of the proposed predicting traffic safety risk model.

the FHWA's *Toward Zero Deaths* vision, one of the challenges researchers and state DOTs face how to identify key contributing factors, and then how to implement effective safety countermeasures accordingly. Safety risks mitigation strategies can only be useful if driver behavior along with external conditions and factors, including weather, roadway conditions, time of day, traffic flow and density, together with their interactions are all accounted. Advanced technology in image processing and Internet of Things can certainly play a significant role in such a holistic risk assessment. The study results confirm that our proposed method is suitable for classifying driver outcomes into different classes. Similar to other studies, this study also has some limitations. In time-series data sets, the surrounding data such as lane marking and radar data are not included. Moreover, time-series variables are not yet available for baseline events. Given this fact, in this study, we present the numerical results of our developed model only for the dichotomous problem of classifying crash and near-crash events. Therefore, the possible extension of this study can focus on, including the time-series variables for baseline events.

9.8 PRACTICAL APPLICATIONS

From our findings, we derive two significant practical implications. First, the identification of driver-related risk factors and outcome can lead to the development of effective strategies to reduce motor vehicle traffic crashes and severity of injuries on the roads. Moreover, it is important for the policymakers, decision-makers, state and local transportation agencies or other related organizations to emphasize their safety campaigns to raise the drivers' awareness of the risky driving behaviors. Second, the classification of driving events has important implications for car insurance companies to develop the application of the behavior-based auto insurance. Nowadays, the usage- and behavior-based auto insurance are considered to be a preferred standard for drivers. Through analyzing the real-time driver data, it is possible to adjust the insurance premium and provide the most competitive premium based on a more accurate picture of a policy member's driving behavior. It should be noted that the behavior-based auto insurance can significantly reduce the number of claims costs and enhance the customer loyalty.

DISCLAIMER

The findings and conclusions of this study are those of the author and do not necessarily represent the views of the VTTI, SHRP 2, the Transportation Research Board, or the National Academies. Furthermore, the contents of this chapter reflect the views of the author, who is responsible for the facts and the accuracy of the information presented herein. The United States Government assumes no liability for the contents or use thereof.

REFERENCES

Ahmed, M., and M. Abdel-Aty. A data fusion framework for real-time risk assessment on freeways. *Transportation Research Part C: Emerging Technologies*, Vol. 26, 2013, pp. 203–213.

Baydogan, M. G., and G. Runger. Learning a symbolic representation for multivariate time series classification. *Data Mining and Knowledge Discovery*, Vol. 29, No. 2, 2015, pp. 400–422.

Das, S., and X. Sun. Association knowledge for fatal run-off-road crashes by multiple correspondence analysis. *IATSS Research*, Vol. 39, No. 2, 2016, pp. 146–155.

Dutta, A., G. Bandopadhyay, and S. Sengupta. Prediction of stock performance in Indian stock market using logistic regression. *International Journal of Business and Information*, Vol. 7, No. 1, 2015, 105–136.

Factor, R., G. Yair, and D. Mahalel. Who by accident? The social morphology of car accidents. *Risk Analysis*, Vol. 30, No. 9, 2010, pp. 1411–1423.

Federal Motor Carrier Safety Administration (FMCSA). *Report to Congress on the Large Truck Crash Causation Study*. Washington, DC, 2006.

Fontaine, H. A. A typological analysis of pedestrian accidents. In *7th Workshop of ICTCT*. Paris, France, 1995.

Gelman, A., and J. Hill. *Data Analysis Using Regression and Multilevel/Hierarchical Models*. Cambridge University Press, Cambridge, UK, 2007.

Hastie, T., R. Tibshirani, and J. Friedman. *The Elements of Statistical Learning: Data Mining, Inference, and Prediction*, 2nd Ed. Springer, New York, 2009.

Hossain, M., and Y. Muromachi. A Bayesian network based framework for real-time crash prediction on the basic freeway segments of urban expressways. *Accident Analysis & Prevention*, Vol. 45, 2012, pp. 373–381.

Jafari, M. A., F. Farzan, K. N. M. N. Al-Khalifa, and T. Gang. Development of a risk-based model using naturalistic driver study. *21st ITS World Congress*, Detroit, Michigan, 2014.

Jalayer, M., and H. Zhou. A multiple correspondence analysis of at-fault motorcycle-involved crashes in Alabama. *Journal of Advanced Transportation*, Vol. 50, 2017, pp. 2089–2099.

Kamei, Y., and E. Shihab. Defect prediction: accomplishments and future challenges. In *Software Analysis, Evolution, and Reengineering (SANER), 2016 IEEE 23rd International Conference on*, Vol. 5, 2016, pp. 33–45.

Kim, K., and E. Yamashita. Corresponding characteristics and circumstances of collision-involved pedestrians in Hawaii. *Transportation Research Record: Journal of the Transportation Research Board*, No. 2073, 2008, pp. 18–24.

Landis, J. R., and G. G. Koch. The measurement of observer agreement for categorical data. *Biometrics*, 1977, pp. 159–174.

Lee, J., and F. Mannering. Impact of roadside features on the frequency and severity of run-off-roadway accidents: an empirical analysis. *Accident Analysis & Prevention*, Vol. 34, No. 2, 2002, pp.149–161.

McGinnis, R., M. Davis, and E. Hathaway. Longitudinal analysis of fatal run-off-road crashes, 1975 to 1997. *Transportation Research Record: Journal of the Transportation Research Board*, No. 1746, 2001, pp. 47–58.

Nallet, N., M. Bernard, and M. Chiron. Individuals taking a French driving licence points recovery course: their attitudes towards violations. *Accident Analysis & Prevention*, Vol. 40, No. 6, 2008, pp. 1836–1843.

National Highway Traffic Safety Administration (NHTSA). (2006). *The 100-Car Naturalistic Driving Study*. National Technical Information Service, Springfield, VA.

National Highway Traffic Safety Administration (NHTSA). *Fatality Analysis Reporting System (FARS)*, Washington, DC., 2015.

National Highway Traffic Safety Administration (NHTSA). *New NHTSA Study Shows Motor Vehicle Crashes Have $871 Billion* Economic *and Societal Impact on U.S. Citizens* (2014), Washington, D.C., 2014. www.nhtsa.gov/press-releases/new-nhtsa-study-shows-motor-vehicle-crashes-have-871-billion-economic-and-societal. Accessed February 15, 2017.

National Highway Traffic Safety Administration (NHTSA). *Traffic Fatalities up Sharply in 2015* (2016), Washington, D.C., www.nhtsa.gov/press-releases/traffic-fatalities-sharply-2015. Accessed February 15, 2017.

Polikar, R. Ensemble learning. *Scholarpedia*, Vol. 4, No. 1, 2008, p. 2776.

Shew, C., A. Pande, A., and C. Nuworsoo. Transferability and robustness of real-time freeway crash risk assessment. *Journal of Safety Research*, Vol. 46, 2013, pp. 83–90.

Suma, V., T. P. Pushphavathi, and V. Ramaswamy. An approach to predict software project success based on random forest classifier. *ICT and Critical Infrastructure: Proceedings of the 48th Annual Convention of Computer Society of India-Vol II* (pp. 329–336). Springer International Publishing, 2014.

Wu, K. F., E. T. Donnell, and J. Aguero-Valverde. Relating crash frequency and severity: evaluating the effectiveness of shoulder rumble strips on reducing fatal and major injury crashes. *Accident Analysis & Prevention*, Vol. 67, 2014, pp. 86–95.

Xu, C., A. P. Tarko, W. Wang, and P. Liu. Predicting crash likelihood and severity on freeways with real-time loop detector data. *Accident Analysis & Prevention*, Vol. 57, 2013, pp. 30–39.

Xu, C., W. Wang, P. Liu, R. Guo, and Z. Li. Using the Bayesian updating approach to improve the spatial and temporal transferability of real-time crash risk prediction models. *Transportation Research Part C: Emerging Technologies*, Vol. 38, 2014, pp. 167–176.

Yang, Y., S. S. Farid, and N. F. Thornhill. Data mining for rapid prediction of facility fit and debottlenecking of biomanufacturing facilities. *Journal of Biotechnology*, Vol. 179, 2014, pp. 17–25.

Yu, R., and M. Abdel-Aty. Multi-level Bayesian analyses for single-and multi-vehicle freeway crashes. *Accident Analysis & Prevention*, Vol. 58, 2013, pp. 97–105.

10 Architecture Design of Internet of Things-Enabled Cloud Platform for Managing the Production of Prefabricated Public Houses

Clyde Zhengdao Li, Bo Yu, Cheng Fan, and Jingke Hong

CONTENTS

10.1 INTRODUCTION

10.1.1 Housing Concerns in Shenzhen and Advantages of Prefabricated Public Houses

Urbanization, with the development of industrialization, is the natural history process of non-agricultural agglomeration in urban and the population shift from rural to urban areas. As a pioneer in China's reform and opening up, Shenzhen takes the lead in achieving urbanization. However, some problems still exist in Shenzhen, one of the most important issues is the balance of housing supply and demand. In the latter part of the accelerated development of urbanization, Shenzhen's population is growing rapidly, which lead to an increase in housing demand. According to the Shenzhen housing construction plan (2016–2020) compiled by Urban Planning, Land & Resources Commission of Shenzhen Municipality, about 1.8 Million housing units are needed during the period between (2016–2020). In the context of high housing price, the Shenzhen government plans to supply more than 400,000 public houses from 2016 to 2020 for easing pressure on the housing market. Meanwhile, Shenzhen is also witnessing a series of dilemmas and constraints, including time, safety, labor shortages, environment protection and quality concerns. Under this socio-economic background and as a solution to housing problems, prefabricated construction comes into the Shenzhen government's sight.

Prefabrication is a manufacturing process that combines various basic materials to form the component parts of the final installation in a specialized facility (Gibb, 1999), which is used to distinguished from traditional construction practice transporting the basic materials to the construction site. Prefabrication construction has long been recognized internationally to have numerous advantages over traditional construction technologies, it has the following benefits: (1) significant savings in construction schedule since concrete maintenance is not needed for components on-site (Tam and Hao 2014); (2) better on-site construction environment as a result of more optimal material usage, recycling, noise capture, and dust capture (Tam, Fung et al., 2015, Hong, Shen et al., 2016); (3) Lower energy consumption, water, and air pollution; (4) more control over the engineering standard (Li, Hong et al., 2016a). Therefore, prefabrication is recommended as a key vehicle in promoting efficient construction as well as in alleviating the adverse environmental impact of conventional cast-in-situ construction within developed construction industries (Li, Shen et al., 2014).

10.1.2 Existing Technical Challenges

Application of prefabricated housing can improve the quality and performance of residential as well as save construction period, providing a safer and cleaner construction environment. However, the construction of prefabricated housing is also faced with a series of technical problems, such as data fragmentation, process discontinuity, poor data interoperability between enterprise information systems (EISs), real-time information lacking visibility and traceability during the process of design, production, transportation, and site assembly (Li, Hong et al., 2016b). In recent years, most of the prefabricated factories in Shenzhen have moved to Dongguan, Huizhou, Zhongshan or other cities in Pearl River Delta region (PRD). Although the labor cost can be reduced, the transport distance of the prefabricated components has been increased. Stakeholders are not in a relatively concentrated area, which is detrimental to the transfer and sharing of information among stakeholders. It also leads to information faults, delays in duration, and increasing costs. As Shenzhen is vigorously promoting the prefabricated buildings, it is imminent to solve this technical problem in order to better manage the construction process.

10.1.3 Internet of Things-Enabled Cloud Platform

The cloud platform is a cloud-based service that provides remote storage services, integration services, identity management services, and data sharing services for all stakeholders. Based on the Internet technology, Internet of Things (IoT) technology is an extended network technology, carrying out information exchange and communication in order to achieve intelligent identification, positioning, tracking, monitoring, and management, it includes building information modeling (BIM), geographic information system (GIS), and Beidou positioning technology. BIM is a digital technology applied to the whole life cycle of prefabricated buildings. It creates, collects, shares, and passes all relevant information about the building and establishes a database of information sharing in a common data format. The BIM technology facilitates the real-time management and decision-making of all stakeholders in the construction of prefabricated public houses. GIS is a kind of technical system which comprehensively processes and analyzes geospatial data. It can quickly read, analyze, and manage geospatial data. Beidou positioning technology can be achieved prefabricated components of the precise positioning of transport vehicles and prefabricated components of the precise assembly. In the stage of logistics transportation and on-site assembly, Beidou and GIS integrated technologies can realize precise positioning and spatial visualization of prefabricated components. With the technical support from IoT, smart city need to have three features of being instrumented, interconnected, and intelligent. Only then a smart city can be formed by integrating all these intelligent features at its advanced stage of IoT development.

This study has developed a cloud-based IoT technology to solve the technical problems in the construction of prefabricated housing. First of all, through the cloud platform, the cutting-edge loT technology can be used to create a smart construction environment, making the smart objects sense and interact data with each other. Secondly, in order to meet the requirements of the three core stages of prefabricated components, the cloud platform provides production management services, logistics management services and on-site assembly management services for prefabricated components. Meanwhile, it collects real-time information on prefabricated components during production, transportation, and on-site assembly, and then feed the information back to the stakeholders according to a predefined workflow. Finally, the stakeholders can share the real-time information generated at all stages through the cloud platform, realizing the interoperability service of the data source. On the whole, different data (including BIM data, Beidou data, and GIS data) are superimposed on the cloud platform, it can realize the data transmission in the whole life cycle of the prefabricated public house. At the same time, the cloud platform can be used as the basis for data exchange and sharing between smart city application systems.

10.2 RESEARCH BACKGROUND

10.2.1 Massive Production of Prefabricated Public Houses in Shenzhen, Advantages of Prefabrication and Policies to Promote Prefabrication

In the recent one year, the Chinese government at all levels had issued a series of measures to further enhance the promotion of prefabricated building (Figure 10.1). The "Several opinions on further strengthening the management of urban planning and construction" proposed by The State Council points out that the proportion of prefabricated buildings to new buildings should reach 30% through 10 years of efforts. In order to achieve this goal. The State Council put forward eight key tasks in "Guidance on the development of prefabricated buildings" in September 2016, including: (1) improving the standard system; (2) innovating the design theory of prefabricated building; (3) optimizing the production of components; (4) enhancing the level of prefabrication construction; (5) promoting collaborative construction; (6) promoting green building materials; (7) implementing general constructing; and (8) improving the quality and safety management system of prefabricated building. In March 2017, the

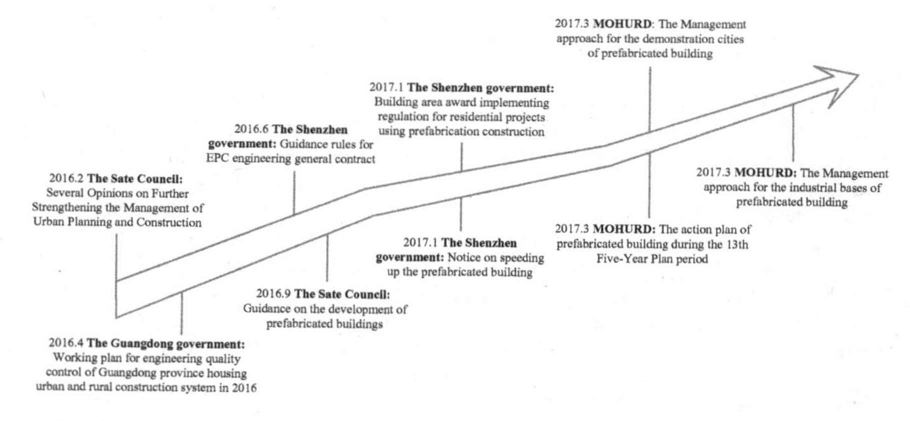

Figure 10.1 The measures issued by Chinese government.

Ministry of Housing and Urban-Rural Development of the People's Republic of China (MOHURD) issued "the management approach for the demonstration cities of prefabricated building," "the action plan of prefabricated building during the 13th Five-Year Plan period," and "the management approach for the industrial bases of prefabricated building," defining the development goals of 2020: (1) the proportion of prefabricated buildings to new buildings should reach at least 15%; and (2) at least 50 demonstration cities, 200 industrial bases for prefabricated building, 500 demonstration projects of prefabricated building and 30 science and technology innovation bases for prefabricated building are cultivated. In response to the country's call for promoting prefabricated building, the Guangdong government issued "Working plan for engineering quality control of Guangdong province housing urban and rural construction system in 2016," indicating that Guangdong province will increase policy support to promote the development of prefabricated building. As an important city in the PRD region and a special economic zone, Shenzhen plays an important role in promoting prefabricated building. The Shenzhen government issued a series of measures, including *"Guidance rules for EPC engineering general contract,"* "Notice on speeding up the prefabricated building," and "Building area award implementing for regulation residential projects using prefabrication construction." According to these measures, prefabrication construction will be supported and encouraged from each process, e.g., bidding, production, construction permit, quality, and safety supervision.

10.2.2 Technical Challenges of Managing Massive Production of Prefabricated Public Houses

As a new type of construction, prefabricated houses involve more stakeholders (such as off-site producers of prefabricated components, third-party logistics companies, and IT providers) than traditional cast-in-place houses (Jia, Sun et al., 2017). Most of the stakeholders do not involve in the full life cycle of the project. As a result, with the construction of prefabricated housing, stakeholders will continuously join or exit the project, causing the fragmentation of the data transfer between various stakeholders. At the same time, in order to maximize their own interests, stakeholders often ignore the indirect losses due to the lack of timeliness in the process of data transfer. Therefore, it is the timely and effective data transfer in the process of prefabricated housing that matters.

Serving as a platform for data sharing, BIM facilitates the data transfer and communication between stakeholders (Figure 10.2). However, a series of technical problems exist regarding data sharing in the process of production, transportation, and site assembly of prefabricated components. The first problem is how to ensure the timeliness of information in the delivery process. The second one is how to record information through advanced information technology. Because the smartphone has the ability to read, input, and protect information, it is used as a tool for information transmission in the construction of prefabricated houses. But there are still some technical problems of applying smartphone to the construction of prefabricated housing. For example, how to improve the specific smartphone devices that are suitable for the construction industry, including readers, quick response code, and communication networks. These components may have special requirements, such as reading distance, precision, and unit costs. At the same time, how to convert common information into standard format information through a smartphone, so that it can be transferred to the BIM database.

In the process of constructing prefabricated housing, each stakeholder develops its own EIS according to its own information needs (Figure 10.3). Meanwhile, they have also developed Enterprise IT System or purchased standardized Enterprise IT System software packages (such as SAP), which play an important role in the actions among various stakeholders. However, due to the differences in databases, functions as well as modes of operation, these systems cannot recognize each other and lack interoperability. In addition, there is a confrontational culture existing in the construction industry, stakeholders often focus on their own interests and seek to maximize their interests, let alone share information (Zhong, Peng et al., 2017). In the process of production, transportation, and site assembly of prefabricated components, stakeholders become increasingly aware of the need for strengthening the interoperability between EISs. Nowadays, BIM is considered as a bridge to connect each EIS, integrating all the information related to the project, including the resources of each stakeholder, the progress of the project and the foreseeable risks. However, the problem of interoperability between BIM and the existing EIS poses new obstacles to the

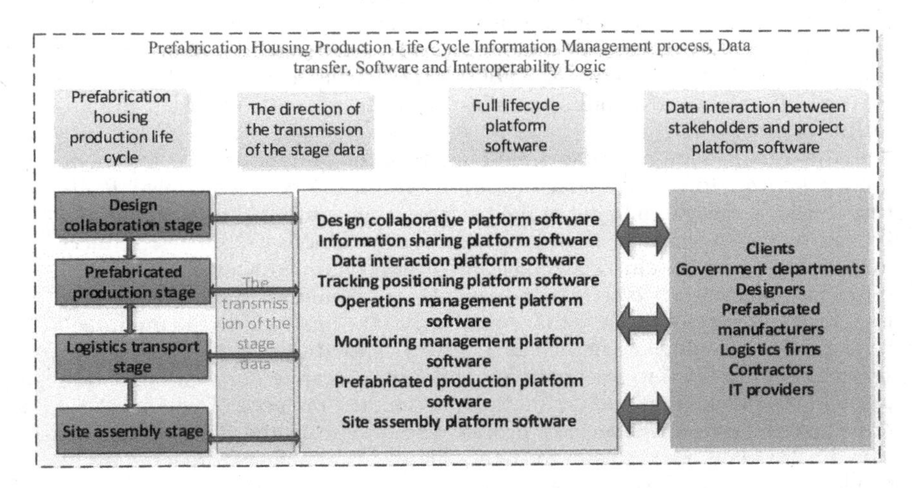

Figure 10.2 Prefabrication Housing Production Life Cycle Information Management process, Data transfer, Software, and Interoperability Logic.

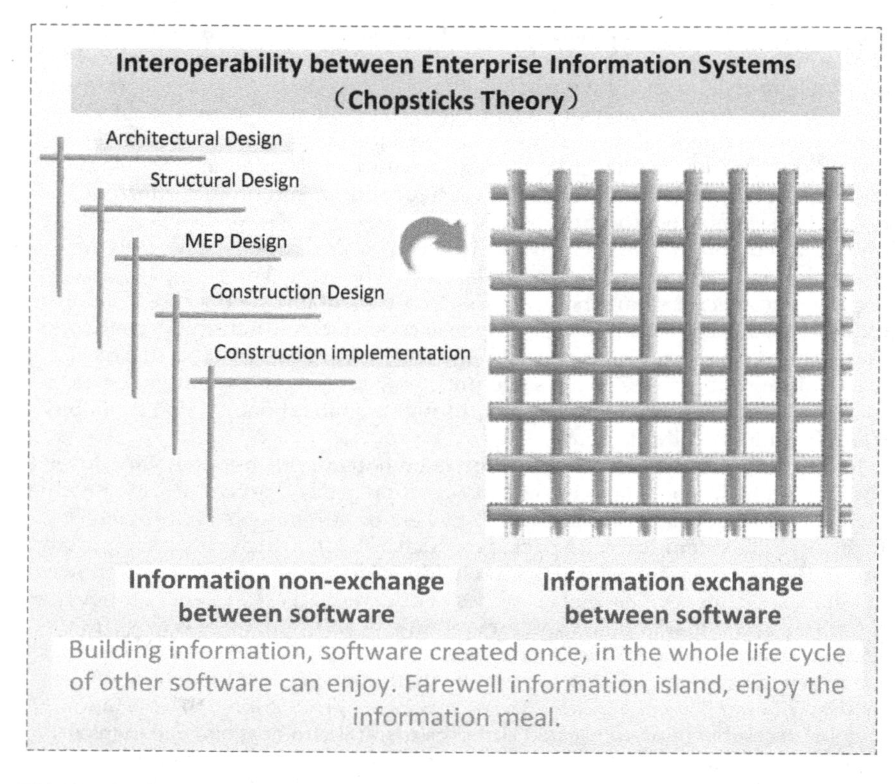

Figure 10.3 Interoperability between Enterprise Information Systems (Chopsticks Theory).

information exchange between various stakeholders. In the on-site assembly stage, there is a lack of data standards for exchanging BIM information and tools for gathering and reading information. It is a big challenge to deliver information from the BIM database to assembly workers. Only to connect BIM and the existing EIS can improve interoperability, which is currently regarded as a huge technical problem.

In the construction of prefabricated housing, the prefabricated components are transported to each construction site by truck and then assembled. Real-time tracking and monitoring of the prefabricated components throughout the entire transport process not only improves productivity, but also reduces interference from the entire logistics and supply chain. However, real-time information of prefabricated components lacks visualization and traceability. Logistics and supply chain management (LSCM) originated in the manufacturing industry, which is defined as a network showing that the products or services are provided to the upstream and downstream enterprises of end-users in the production and circulation process, the purpose of which is to minimize the time spent on each process and maximize the value of each echelon. LSCM plays a key role in the overall logistics management, depending to a large extent on the accuracy and timeliness of information sharing among stakeholders because of the need for industrial production of prefabricated components.

In order to improve the accuracy and timeliness of information, real-time information of prefabricated components can be obtained through the BIM platform, and characterized by visualization and traceability. BIM is not yet able to synchronize the geographic information and location information of prefabricated components through techniques such as smartphone, GIS, and Beidou, so it is an "invisible BIM." The inaccurate and out-of-date data generated from the previous stage can be re-entered into a separate system like BIM and EIS, but these independent systems cannot optimize LSCM. Due to the particularity and massive amount of the prefabricated components, it is necessary to install the corresponding prefabricated components in the exact position within the specified duration period, which is difficult to achieve in China, especially considering the shortage of resources, limited space, and time constraints in Shenzhen. So, it is necessary to put forward practical solutions. For example, seeking information sharing systems based on IoT technology to coordinate project stakeholders to better achieve LSCM. Therefore, based on this special situation, it is urgent to combine it with BIM to build a platform that can form just-in-time (JIT) distribution system to improve the efficiency of prefabricated components in logistics management.

10.3 ARCHITECTURE DESIGN FOR INTERNET OF THINGS-ENABLED CLOUD PLATFORM

10.3.1 Processes and Function Requirements

The prefabrication components for housing development consist of three critical stages: prefabrication production, logistics, and on-site assembly. There are four main steps for the production of prefabricated components including receiving order for prefabrication production, preparation for prefabrication production, prefabrication production, and prefabrication logistics preparation. The prefab logistics preparation links to logistics stage. Based on the project master plan, the prefabrication manufacturer would devise a corresponding prefabrication production plan to ensure the project progress is on schedule. Once the final products are ready, the logistics company would deliver the prefabricated components from the component manufacturer to the construction site in Shenzhen. This logistics requires the visibility of logistics status and real-time information in order to achieve JIT delivery. On-site assembly of the prefabrication components is the last stage covered in this study for prefabricated public housing construction. It occurs after the prefabrication components are delivered through the logistics to the construction site. Once the prefabricated components arrive on-site, an on-site foreman or an assigned operator would check for defects/quality before erection. The results and relevant product information will be recorded and stored. The on-site assembly stage can generally be divided into five main stages, namely site establishment, temporary works, superstructure works, architectural works and building services installations.

Through appreciating the latest innovations and technologies in the construction industry, this research develops a cloud platform is designed by integrating big data analytics systems with cloud computing and IoT technologies to manage the massive production of prefabricated public houses in Shenzhen. Firstly, smartphone-enabled multi-dimensional BIM platform with IoT-enabled real-time visibility and traceability for prefabricated construction industry. Problems and technical challenges in prefabrication construction industry in Shenzhen are investigated and identified. An innovative smartphone-enabled real-time BIM platform has been established to integrate people, prefabrication processes, information flow, and technologies. Secondly, the design BIM models approved by the Contract Manager would normally include detailed design information such as prefabrication and in-situ information. The positioning of where the quick response code would be inserted should also be indicated in the design BIM models for standardized prefabrication components. Since the design BIM models are developed by professionals from different disciplines (architectural, structural, site formation, building services, and landscaping), the contractor is required to develop drawings and BIM models for construction as well as to facilitate planning and design coordination, in order to minimize repetitive works and waste generation. Thirdly, by describing the complex spatial data and attribute data through the form of map, GIS analyzes the life cycle of a fabricated building, which can realize the visualization of the combination of text and image. Meanwhile, GIS system can acquire, analyze, and manage geospatial data, and then quickly read them. By combining GIS and Life Cycle Assembly, abstract data is linked to temporal and spatial locations. As a result, stakeholders can effectively master the specific location of prefabricated components in real-time and visualize complex spatial and attribute data. Through the integration of intelligent components, the flow diagram of the fabricated building in their whole life cycle can be drawn, which achieves a combination of text, images and information on positioning, query retrieval. In general, the cloud platform based on the IoT technology is to effectively manage a large number of data generated by BIM, smartphone, and GIS, facilitating data interaction and sharing among various stakeholders and various stages of fabricated building in their life cycle. The proposed enhancements to prefabrication logistics service, as part of the proposed solution, aims to achieve real-time visibility of logistics status and JIT logistics operations to provide seamless linkage and delivery tracking between prefabrication factories and the construction sites.

10.3.2 Service-Oriented Architecture Design

A complete service-oriented architecture process involves three main stages: Publish, Search, and Invoke (Figure 10.4), (Capretz, Capretz et al., 2016; Li, Zhong et al., 2017). Service developers/providers deploy their web services at

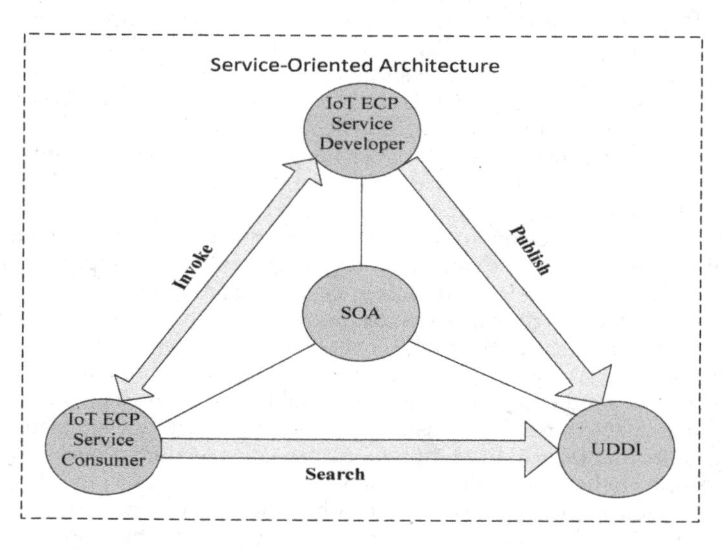

Figure 10.4 Service-Oriented Architecture for IoTECP.

the designated server sites and publish the details of these services including location, capability as well as interfacial description. Service consumers search and select appropriate web services from the published database. Values must be specified and provided before services are invoked during a specific application process.

These three typical stages involve three fundamental web services tools. They are universal description, discovery and integration (UDDI), web services description language (WSDL), and simple object access protocol (SOAP). UDDI is a platform-independent, XML-based registry for distributed services to list themselves on the Internet, while WSDL standard provides a uniform way of describing the abstract interface and protocol bindings of these services. In simple terms, a WSDL describe what a web service can do, where it resides, and how to invoke it. SOAP is a platform-independent protocol for invoking those distributed web services through exchanging XML-based messages.

10.3.3 Gateway Operation System and Quick Response Code for Defining Intelligent Building Elements

IoTECP for production of prefabricated public houses gateway uses an operating system named gateway operation system (GOS) to achieve a flexible, modularized, and re-configurable framework, where applications and solutions are designed and developed as web services. The overall architecture designs of GOS shown in Figure 10.5. Focuses will be placed upon four sections to be constructed in the GOS, namely gateway device library, gateway smart agent manager, gateway application information service and gateway application manager (Fang, Qu et al., 2013). GOS is an overall middleware solution, which is designed and proposed not only to address basic functions of smartphones, but also to overcome the particular challenges in real-life prefabrication manufacturing progresses. It aims to provide an easy-to-deploy, simple-to-use and flexible-to-access solution for the construction industry. Within GOS, multi-agent-based models are used to ensure the versatility and scalability of IoTECP gateway. Therefore, communication and interactions between Smart Construction

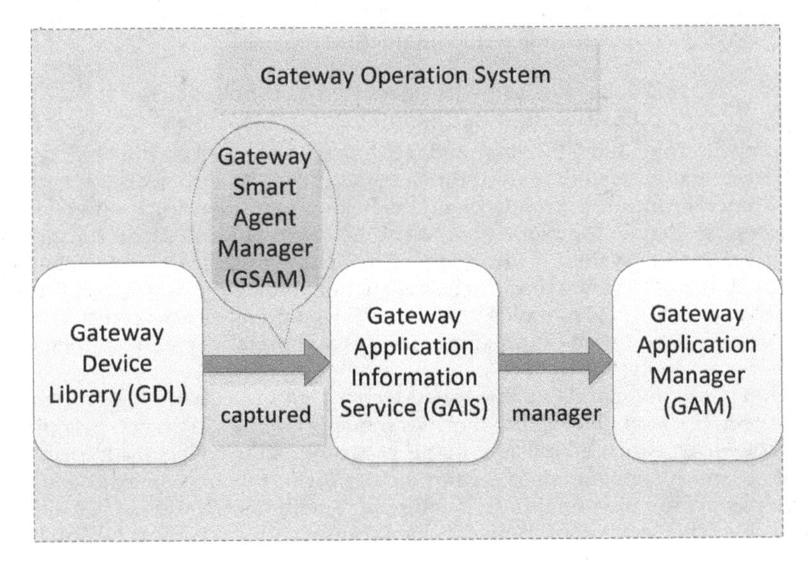

Figure 10.5 Gateway Operation System.

Objects (SCOs) and other services is facilitated by using an XML-based message exchanging protocol.

Quick response (QR) code is a certain geometric graphics in accordance with certain laws in the plane (two-dimensional direction) distribution of black and white graphics data symbolic information; in the code to make use of the computer composition of the internal logic of the "0," the concept of "1" bit stream, the use of a number of binary corresponding to the geometric shape to represent the text value information, through the image input device or photoelectric scanning device automatically read to achieve automatic information processing.

By scanning QR code with the smartphone, the relevant information of the prefabricated component can be obtained and the production, logistics, and on-site assembly stage of the prefabricated component can be monitored in real-time. In the prefabricated production stage, the basic information of each prefabricated component is written into the QR code, and the QR code is bound to the surface of the prefabricated component. The QR code mainly records the basic information such as manufacturer, production date, specification, material composition, and product quality inspection record. In the logistics transportation stage, the workers scan the QR codes of prefabricated components with smartphones and check whether the basic factory information of prefabricated components is consistent with the delivery order. If consistent, the transport worker will write the transport information, reasonably plan the transportation sequence, generate the transport route, select the transport vehicle, and upload the information of the transport vehicle to the BIM database. After the prefabricated components are assembled, the workers scan the QR codes with smartphones and upload progress and quality information of prefabricated components into the BIM database. For some stakeholders like enterprise, prefabrication manufacturers, logistics partners, and designers, it is possible to judge the assembly progress of prefabricated components and check whether the assembly quality conforms to the construction specifications and technical requirements through the BIM database. For other interested stakeholders like investment, statistical reports will be generated according to the data captured from front-line construction sites through the BIM database. Thus, all stakeholders are able to make decisions in real-time based on the BIM database.

10.3.4 Beidou and GIS Integrated Technologies for Positioning Intelligent Building Elements

Beidou positioning and GIS integrated technologies is based on the Beidou satellite navigation system for real-time positioning of space objects advanced technology (Figure 10.6). Among them, the Beidou satellite navigation system is China's independent implementation and has been put into use in the global satellite navigation system, with advanced technology, stable operation characteristics. In the assembly of building construction management process, through the Beidou positioning technology can be achieved prefabricated components of the precise positioning of transport vehicles and prefabricated components of the precise assembly.

Beidou positioning and GIS integrated technologies can be divided into three stages: prefabricated production, logistics transportation, and on-site assembly. First, download a smartphone app that is programmed to collect the information about production and delivery through the App store on a smartphone. The first step is to bind the prefabricated component with the QR code. After the code is fixed on the reinforcement cage of a component, the worker will open the "binding" page in the smartphone App. In that page, a list of components from the database is arranged by the planned production date. The worker will

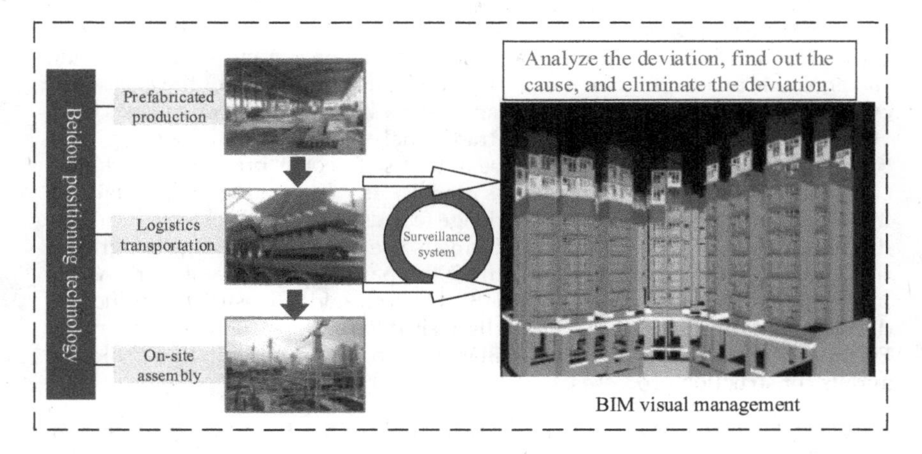

Figure 10.6 Beidou positioning technology and BIM visual management.

select the right one and scan the QR code by the App. In this way, the code will be linked with the component that it attaches to. Meanwhile, the binding date and Beidou data will be automatically collected by the App and uploaded to the database. Second, the purpose of the real-time transportation monitoring is to get the current statuses and locations of the prefabrication components throughout the logistics and supply chain. It uses smartphone and Beidou technologies to real-time trace and track the transportation vehicles through visualizing the data, which finally could be converted into meaningful information and finally turned into 3D graphical presentation of the prefabrication logistics. Through the graphical presentation, critical prefabrication logistics statuses, progress, as well as current locations will be real-time reflected by the smartphone and Beidou data. Finally, Cumulative quantity of precast elements erected based on real-time data collected and based on contractor's master programme can be compared in line chart to identify any delay in site construction progress. It provides a Gantt Chart or a 3D virtual reality presentation which uses the smartphone assembly data to real-timely reflect the construction progress in terms of prefabrication assembly status, material consumptions, and workers assignment. The main users are on-site supervisors who are responsible for controlling the construction objects and reporting to various stakeholders about the progress, current challenges, or barriers.

10.3.5 Dynamic nD BIM for Visualizing the Whole Processes of Prefabrication Construction Processes

According to Eastman (1999), BIM is "a digital representation of the building process to facilitate exchange and interoperability of information in digital format." BIM can be used for a wide range of purposes, e.g., design and construction integration, project management, facilities management. BIM has a direct impact on project performance. With its 3D presentation and virtual reality simulation capability, BIM was introduced as a technical tool that can be applied to improve productivity in a number of areas, such as improving design quality, construction plan rehearsal and optimization, and construction site management (Kaner, Sacks et al., 2008, Li, Lu et al., 2009). BIM is argued to be a useful tool for reducing the problems in the construction industry, such as fragmentation, improving its efficiency/effectiveness, and lowering the high costs of inadequate interoperability. BIM is also recognized as a virtual design and construction (VDC) environment, a vehicle facilitating communications among stakeholders, an information model that can be used throughout the project life cycle, or an education platform that can be used in universities or colleges (Lu, Huang et al., 2011). It is changing the traditional architecture, engineering, and construction practices in a broad sense in terms of people, processes, working culture, communication, and business models. As a continuing digital platform, it can retain the information or knowledge (e.g. design rationale) to reduce the discontinuity (Li, Lu et al., 2009). BIM can also be used to encourage integration and collaboration (Taylor and John, 2009), particularly when it works with the integrated project delivery model. BIM is even said to be such a significant development that it will bring a paradigm shift to the construction industry. Figure 10.7 is a typical screenshot of BIM used to represent digitally a building and its construction processes.

10.3.6 Data Source Management Service through Big Data Analytics Systems

Big data analysis system can be divided into six processes: data acquisition, data storage, data calculation, data mining, data presentation, and data security, in which the data are mainly depending on the stakeholders' behavior, the construction state and so on (Figure 10.8). In order to achieve real-time data, smartphone, gateway system as well as visualization and traceability system play an important role. SCOs are typical prefabricated components of intelligent factors. It can be construction tools, machinery, materials, and other information binding to different smartphone devices, and then converted into intelligent components. SCOs are designed to create an intelligent building environment for prefabricated components (such as production plants, warehouses, transport processes, transit stations, and construction sites). The IoTECP gateway connects all the SCOs to collect real-time information on prefabricated components during production, transportation, and field assembly. The visualization and traceability system can obtain real-time data from a smartphone and the data obtained are stored in the IoTECP. Data

Figure 10.7 A screenshot of BIM used to represent digitally a building and its construction processes.

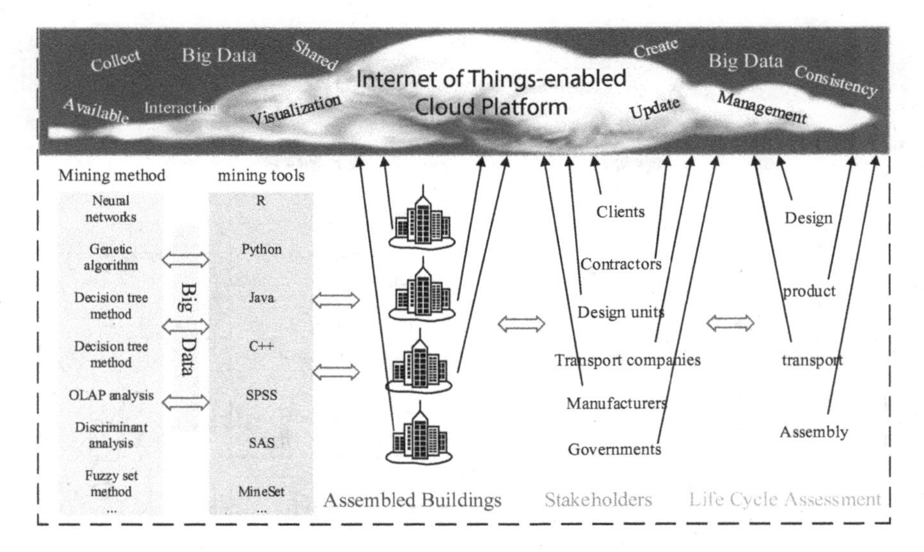

Figure 10.8 Internet of Things-enabled Cloud Platform.

source interoperability services integrate EIS for data sharing and improving system interoperability, which can complete their own tasks in the use of agency technology without human intervention. Meanwhile, different data sources (such as communication protocols, information presentations) with high heterogeneity and unformatted EDI (Electronic Data Interchange) are standardized to facilitate data mining. According to the calculation results, the decision support system provides the decision-making service for the decision-makers in the prefabricated production stage, the logistics transportation stage and site assembly stage.

10.4 BENEFITS TO STAKEHOLDERS INVOLVED IN MASSIVE PRODUCTION OF PREFABRICATED PUBLIC HOUSES

The developed IoTECP is available to all stakeholders including clients, contractors, prefabrication manufacturers, third-party logistics firms, and IT vendors (Figure 10.9). Using the IoTECP, stakeholders can not only improve productivity and quality of housing, reduce the waste of construction resources, transportation costs and potential risks, but also increase the number of housing supply, to solve the problem about the current housing shortage in Shenzhen.

10.4.1 Benefits and Marketing Opportunities for Clients

IoTECP creates an efficient extended ICT environment for BIM implementation in housing prefabrication production in that the clients can obtain potential benefits of prefabrication and BIM technology in terms of time, cost, and quality from improved project management. Moreover, IoTECP creates a user-friendly and customer-oriented platform for clients to maintain timely and visual supervision on the project progress by IoTECP data source interoperability service (IoTECP-DSIS) and enhance their involvement in the project decision-making processes by IoTECP decision support system (IoTECP-DSS). Clients can monitor the progress of their projects and examine the value created throughout the project's life cycle. It may eventually benefit the general public in Shenzhen as end users of housing.

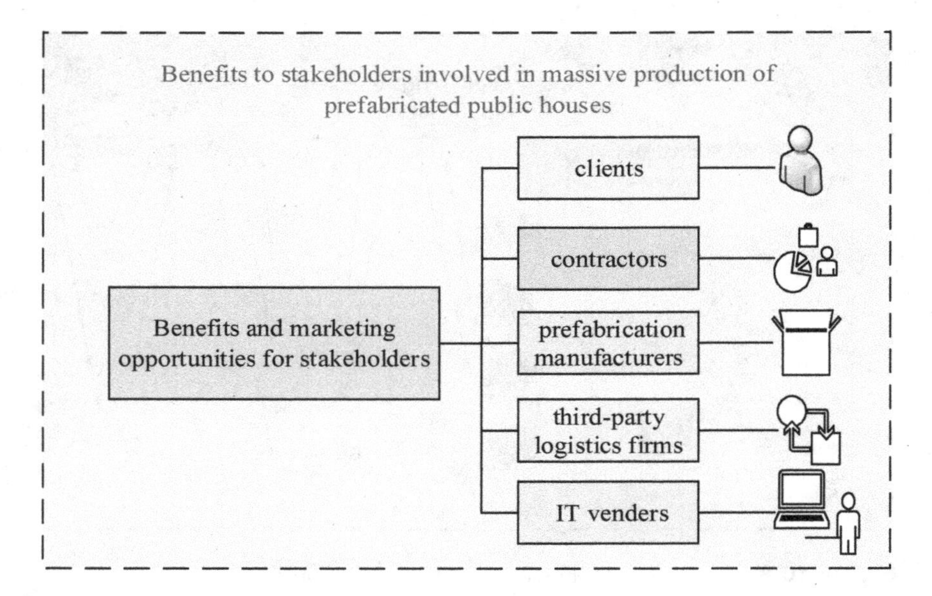

Figure 10.9 Benefits to stakeholders involved in massive production of prefabrication public houses.

10.4.2 Benefits and Marketing Opportunities for Contractors

The marketing opportunities of IoTECP for contractors are significant in many aspects. First, IoTECP increases the trackability, traceability and visualization of project resources, such as labor, materials and equipment by smartphone-enabled SCOs and BIM gateways. Second, IoTECP improves real-time information transfer efficiency among BIM system and on-site works in that the monitoring, control, and management of on-site construction work can be well carried out, and the working environment for construction workers can be safe and healthy. Third, IoTECP-DSIS provides a series of services from prefabrication production, to prefabrication transportation and on-site assembly, which could result in the significant increase in project delivery efficiency. Fourth, IoTECP provides a good solution for interoperability issue among heterogeneous EISs of project participants through IoTECP-DSS. Therefore, the communication efficiency among contractors and other stakeholders, such as clients, architects, consultants, engineers, suppliers, and vendors could be improved, thereby resulting in improved time and cost control, less rework and disputes, and high productivity.

10.4.3 Benefits and Marketing Opportunities for Prefabrication Manufacturers

IoTECP-DSS could provide a production service module to prefabrication manufacturers on production planning, production scheduling, internal logistics, and production execution. First, production orders could be generated from a BIM system by selecting and translating to-be-prefabricated components into the required format for prefabrication manufacturers. Second, production scheduling and execution could be well conducted with the assistance of the information technology within IoTECP. Third, the internal logistics of prefabrication production could be better tracked and managed by embedding QR code. Fourth, the produced prefabrication components can be well organized for storage, transportation and assembly. Moreover, IoTECP-DSS could establish ordering standards, production principles and processes and manage prefabrication logistics for the prefabrication industry. The prefabrication components used in Shenzhen housing projects are imported mainly from prefabrication manufacturers in PRD. Shenzhen-based industrial standards of prefabrication production could provide greater convenience and result in a "spill over" effect.

10.4.4 Benefits and Marketing Opportunities for Third-Party Logistics Firms

Significant marketing demands for third-party logistics companies across the PRD-Shenzhen border have resulted from the special situation that Shenzhen housing projects demand a huge amount of prefabrication components imported from prefabrication firms in PRD. IoTECP can provide IT-based prefabrication transportation services on transportation planning and scheduling and fleet management that could help third-party logistics companies streamline logistics management processes. However, by employing smartphone and Beidou technologies, IoTECP can realize real-time monitoring and execute cross-border logistics efficiently. For example, IoTECP can make the customs clearance more convenient for both customs and third-party logistics companies because the QR code attached to prefabrication components could be well recognized by customs' electronic custom clearance system.

10.4.5 Benefits and Marketing Opportunities for IT Vendors

Skilled IT professionals are needed to operate and maintain IoTECP. First, IoTECP could improve the housing prefabrication production in Shenzhen with the thrust from the Shenzhen Talents Housing Group because of the designed interface between Shenzhen Talents Housing Group's (STHG) enterprise system and IoTECP. With necessary training, IoTECP could be well hosted by STHG IT professionals to enhance enterprise system functions. Second, IT companies that promote BIM/VDC technologies in the construction industry can find considerable opportunities in IoTECP. These IT companies can become more competitive in the ICT market because of the effective interface of IoTECP with BIM system.

10.5 DISCUSSION AND FUTURE RESEARCH

With the transformation and upgrading of China's economic and social development, and vigorously develop the assembly-style architecture at the right time, the IoT technology in the construction and management play an increasingly important role. This paper puts forward a cloud platform for the management of the fabricated public housing in Shenzhen. In the production of fabricated buildings, using the smartphone technology to solve the lack of information processing of intelligent building components and the lack of precision assembly process, the use of Beidou and GIS integrated technologies for positioning intelligent building elements, dynamic nD BIM for visualizing the whole processes of prefabrication construction processes to solve the problem of visual management, based on cloud technology on the city large-scale assembly of large data interoperability and management. Through the use of BIM, smartphone, GIS and Beidou and other visualization techniques, the establishment of assembly-building life-cycle information management system, to achieve the whole process of quality control and management to provide information support, to further promote the development of assembly-style buildings. The IoTECP provides the ability to remotely monitor, manage and control devices, and to create new insights and actionable information from massive streams of real-time data for smart city construction.

Although the platform has the above advantages, but it does not take into account the actual situation of the workers, so it should be perfected from the following aspects in the future research. First, because most of the workers engaged in prefabricated production in assembly buildings have not received good education and are not aware of the power of high-tech IT technology, and thus to a certain extent, will reduce the implementation of the platform effect. Therefore, technical training courses should be arranged to demonstrate the main functions of the platform and to help workers to become familiar with the specific operation. Second, the user interface of the network system and the smartphone application should be improved so that the user interface is easy to understand and should be consistent with the worker's operating habits. Finally, in future studies, error correction techniques should be introduced to eliminate errors caused by operations.

REFERENCES

Capretz, M. A. M., M. A. M. Capretz, and M. Perry (2016). Trust-Based Service-Oriented Architecture, Elsevier Science Inc, New York, NY.

Fang, J., T. Qu, Z. Li, G. Xu, and G. Q. Huang (2013). "Agent-based gateway operating system for RFID-enabled ubiquitous manufacturing enterprise." Robotics and Computer-Integrated Manufacturing **29**(4): 222–231.

Gibb, A. G. (1999). Off-Site Fabrication: Prefabrication, Pre-Assembly and Modularisation, Wiley, London, UK.

Hong, J., G. Q. Shen, C. Mao, Z. Li, and K. Li (2016). "Life-cycle energy analysis of prefabricated building components: an input–output-based hybrid model." Journal of Cleaner Production **112**: 2198–2207.

Jia, J., J. Sun, Z. Wang, and T. Xu (2017). "The construction of BIM application value system for residential buildings' design stage in China based on traditional DBB mode." Procedia Engineering **180**: 851–858.

Kaner, I., R. Sacks, W. Kassian, and T. Quitt (2008). "Case studies of BIM adoption for precast concrete design by mid-sized structural engineering firms." Electronic Journal of Information Technology in Construction **13**: 303–323.

Li, C. Z., J. Hong, F. Xue, G. Q. Shen, X. Xu, and L. Luo (2016a). "SWOT analysis and Internet of Things-enabled platform for prefabrication housing production in Hong Kong." Habitat International **57**: 74–87.

Li, C. Z., J. Hong, F. Xue, G. Q. Shen, X. Xu, and M. K. Mok (2016b). "Schedule risks in prefabrication housing production in Hong Kong: a social network analysis." Journal of Cleaner Production **134**: 482–494.

Li, C. Z., R. Y. Zhong, F. Xue, G. Xu, K. Chen, G. G. Huang, and G. Q. Shen (2017). "Integrating RFID and BIM technologies for mitigating risks and improving schedule performance of prefabricated house construction." Journal of Cleaner Production, 165: 1048–1062.

Li, H., W. Lu and T. Huang (2009). "Rethinking project management and exploring virtual design and construction as a potential solution." Construction Management & Economics **27**(4): 363–371.

Li, Z., G. Q. Shen, and X. Xue (2014). "Critical review of the research on the management of prefabricated construction." Habitat International **43**(3): 240–249.

Lu, W., G. Q. Huang, and H. Li (2011). "Scenarios for applying RFID technology in construction project management." Automation in Construction **20**(2): 101–106.

Tam, V. W. Y., I. W. H. Fung, M. C. P. Sing, and S. O. Ogunlana (2015). "Best practice of prefabrication implementation in the Hong Kong public and private sectors." Journal of Cleaner Production **109**: 216–231.

Tam, V. W. Y. and J. J. L. Hao (2014). "Prefabrication as a mean of minimizing construction waste on site." International Journal of Construction Management **14**(2): 113–121.

Taylor, E. J. and F. P. Bernstein (2009). "Paradigm trajectories of building information modeling practice in project networks." Journal of Management in Engineering **25**(2): 69–76.

Zhong, R. Y., Y. Peng, F. Xue, J. Fang, W. Zou, H. Luo, S. T. Ng, W. Lu, G. Q. P. Shen, and G. Q. Huang (2017). "Prefabricated construction enabled by the Internet-of-Things." Automation in Construction **76**: 59–70.

Index

A

B

C

D

E

F

G

I

L

M

N

P

R

S